SPEED READ
MUSTANG

Brimming with creative inspiration, how-to projects, and useful information to enrich your everyday life, Quarto Knows is a favorite destination for those pursuing their interests and passions. Visit our site and dig deeper with our books into your area of interest: Quarto Creates, Quarto Cooks, Quarto Homes, Quarto Lives, Quarto Drives, Quarto Explores, Quarto Gifts, or Quarto Kids.

Inspiring | Educating | Creating | Entertaining

© 2018 Quarto Publishing Group USA Inc. Text © 2018 Donald Farr

First published in 2018 by Motorbooks, an imprint of The Quarto Group, 401 Second Avenue North, Suite 310, Minneapolis, MN 55401 USA. T (612) 344-8100 F (612) 344-8692 www.QuartoKnows.com

Motorbooks titles are also available at discount for retail, wholesale, promotional, and bulk purchase. For details, contact the Special Sales Manager by email at specialsales@quarto.com or by mail at The Quarto Group, Attn: Special Sales Manager, 401 Second Avenue North, Suite 310, Minneapolis, MN 55401 USA.

10 9 8 7 6 5 4 3 2 1

ISBN: 978-0-7603- 6042-2

Library of Congress Control Number: 2017963137

Acquiring Editor: Darwin Holmstrom
Project Manager: Alyssa Lochner
Art Director: James Kegley
Cover and page design: Laura Drew

Cover and interior illustrations by Jeremy Kramer.

Printed in China

SPEED READ
MUSTANG

THE HISTORY, DESIGN AND CULTURE BEHIND FORD'S ORIGINAL PONY CAR

DONALD FARR

INTRODUCTION

Standing at a podium within the Ford Pavilion at the New York World's Fair, Lee Iacocca spoke like a proud father as he introduced the new Ford Mustang to the world's press reporters on April 13, 1964. "We think people will want the Mustang because it offers them a 'different' kind of car at low cost," Iacocca said, "because it satisfies their need for basic transportation and their desire for comfort, fresh style, good handling, and a choice of performance capabilities. This is the car we have designed with young America in mind."

For Iacocca, everything was on the line—his reputation, his legacy, even his career. Ford Motor Company CEO Henry Ford II, still reeling from the Edsel failure, had made it clear when he approved the budget to develop Iacocca's new model: "You've got to sell it, and it's your ass if you don't!"

In hindsight, we now know that Iacocca's job—and successful future— was safe. Bolstered by a creative (and expensive) marketing campaign, Ford dealers sold 22,000 Mustangs during the first on-sale weekend, over 120,000 by the end of summer 1964, and more than 680,000 before the 1965 model year ended. Sales topped 1 million by February 1966, making the Mustang the most successful vehicle launch in American automotive history.

Although based on the Falcon, the Mustang's combination of sporty styling, low cost, long list of options, and practicality as a four-seater with a trunk transformed Ford's economy compact into a trendsetter, one that spawned an entirely new "Pony Car" segment—named after the Mustang, of course. Soon, the American highways were filled with Mustangs, Camaros, Cougars, Firebirds, Barracudas, and Challengers.

But unlike its upstart competitors, Mustang production has continued uninterrupted for over fifty-three years, something no other American vehicle nameplate can claim—not even the Corvette, which was introduced eleven years before the Mustang but skipped the 1983 model year entirely.

For millions of owners worldwide, the Mustang is more than a car. It's also a lifestyle, one supported by clubs, parts manufacturers and suppliers, magazines and websites, and specific model registries. The Mustang Club of America alone boasts over one hundred regional clubs, not only in the United States but also worldwide in countries like France, Italy, Taiwan, and Brazil. These enthusiast organizations host thousands of shows, cruises, races, and rallies each year, bringing owners together to celebrate their common allegiance to the Mustang. Few other brands, automotive or otherwise, can claim that kind of loyalty and devotion.

Over the years, popular models like Mach 1, Boss, Grande, and Shelby have expanded the Mustang's reach. The Mustang has become so ingrained in American culture that it was selected in 1999 for a US Post Office "Celebrate the Century" stamp alongside the Woodstock Music Festival and man walking on the moon. Mustangs have appeared in over five hundred movies, including starring roles in *Gone in 60 Seconds* and *Bullitt*, featuring actor Steve McQueen and a Highland Green fastback that inspired special Bullitt-edition Mustangs from Ford. The Mustang's free-wheeling and fun attitude has made it a popular subject for songs, topped by Wilson Pickett's "Mustang Sally," to this day a popular sing-along tune for dance band and DJ audiences.

Over half a decade since Iacocca stood at that World's Fair podium, the Mustang is still going strong. It has survived pony car competition, two major fuel crises, economic downturns, and even an unsuccessful effort by some within Ford to abandon the traditional rear-wheel-drive for a radical switch to front-wheel-drive based on a Japanese chassis. Through it all, the Mustang has remained true to its original objective as conceived by Iacocca and his team in the early 1960s—fun transportation with style, comfort, good handling, performance, and practicality.

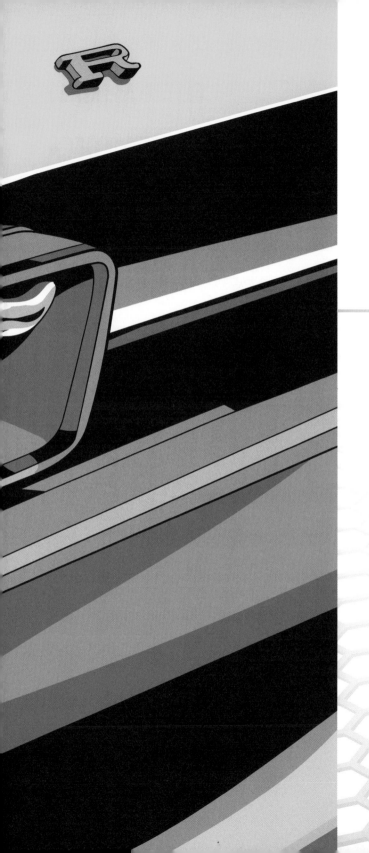

THE LAUNCH

THE LAUNCH
LEE IACOCCA

The son of Italian immigrants, Lee Iacocca went to work for Ford in 1946, starting with a sales job at a Pennsylvania assembly plant. But he had a much bigger career goal—he planned to reach company vice president by the age of 35.

Iacocca strategically worked his way up the ladder, moving into a sales manager position for the East Coast, then assistant district manager. His clever "$56-a-month for a '56 Ford" marketing program led to a promotion and relocation to Dearborn to manage Ford's truck marketing, resulting in record truck sales and yet another promotion into car marketing. In November 1960, Henry Ford II promoted Iacocca to vice president and general manager of Ford Division. Iacocca was 36.

As the top manager of Ford, Iacocca had the horsepower to pursue a hunch, shared by others at Ford, that a new vehicle would appeal to the Baby Boomer generation as they came of age in the mid-1960s. And he knew very well that Chevrolet was selling more of its Corvair, originally an economy car, by offering a sportier Monza model with bucket seats, stick shift, and upgraded interior trim. There was nothing like it in the Ford lineup.

Iacocca realized that economy-based Falcons and boxy Fairlanes weren't the answer. He predicted that, as Baby Boomers matured and reached driving age during the 1960s, the huge youth market would crave sports-car styling and performance combined with the practicality of four seats and a usable trunk. Selling the concept to company President Henry Ford II, who was still reeling from 1958–60 Edsel failure, would be the challenge.

To convince the man whose name was on the building, Iacocca needed a team to help him gather conclusive data. And they had to do it without anyone finding out.

THE LAUNCH
FAIRLANE COMMITTEE

To convince Henry Ford II that Ford needed a new car for the upcoming Baby Boomer generation, Lee Iacocca pulled together a hand-selected think tank of Ford managers representing engineering, styling, product planning, market research, racing, public relations, and advertising. Iacocca would lead the group, which would initially meet twice a month in an attempt to identify a market that could lead to a concept for a new vehicle. To avoid alerting Henry Ford II, the first meetings were held in a private conference room at the Fairlane Inn, a hotel on Michigan Avenue in Dearborn. The clandestine group became known as the "Fairlane Committee."

Ford car marketing manager Chase Morsey was assigned to market research. His digging confirmed that the oldest Baby Boomers would reach car-buying age in the mid-1960s, a time when more than half of projected new-car sales would be purchased by buyers between the ages of 18 and 34. Morsey's surveys also determined that this youthful segment was not intrigued by the traditional styling favored by their parents. They wanted bucket seats, four-on-the-floor shifting, and styling that represented a certain image. Additionally, a booming economy meant that families could add a second car for the wife or teen driver.

Time was of the essence. Noted Iacocca, "We'd hit on such a good thing that we had to get moving before somebody else could come along and beat us to it!"

By late 1961, the Fairlane Committee had determined that there was indeed a market for a new Ford car. They established a list of goals: four-passenger with a sizable trunk, 2,500-pound weight limit, retail price under $2,500, long hood and short rear deck styling, one basic car with many available options, and a target introduction date of April 1964 at the New York World's Fair.

THE LAUNCH
DESIGN CONTEST

Bolstered by the Fairlane Committee's findings, Iacocca pushed forward with the idea of a sporty compact. He realized that developing a totally new car was an expensive proposition, at the time costing upwards of $400 million, a scenario that would be surely rejected by Henry Ford II. Special Projects Assistant Hal Sperlich came up with the solution—build the new car on the Falcon's already existing chassis, drivetrain, and suspension to save both time and money. It was a brilliant no-brainer.

But before Iacocca could approach Henry Ford II with an official proposal, he needed a design, something in clay, to plead his case to make the investment into a new sporty compact. But time was running out. Less than twenty-eight months remained before the target introduction date of April 1964.

During the first half of 1962, Iacocca viewed no less than eighteen clay models. None impressed. So he initiated a design competition between the three Ford Motor Company styling studios—Ford, Lincoln-Mercury, and Advance Projects. On August 16, 1962, Iacocca reviewed six clay models and was immediately drawn to a design from Joe Oros' Ford Studio that included a Ferrari-like grille opening, tri-lens taillights, and side sculpturing that implied rear brake cooling scoops. "One thing hit me instantly," Iacocca said later. "Although it was just sitting there, the brown clay model looked like it was moving."

When Iacocca gambled by inviting Henry Ford II to the styling courtyard for a look, the boss was enthused but not overly excited. "I'll approve the damn thing," HFII reportedly said. " But once I approve it, you've got to sell it, and it's your ass if you don't!"

Henry Ford II officially approved the project on September 10, 1962. Iacocca had only eighteen months to take his hunch from concept to showroom.

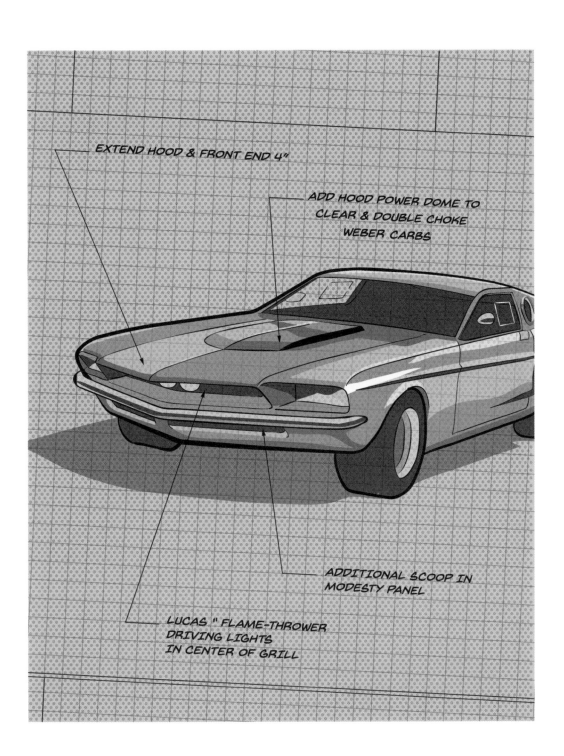

EXTEND HOOD & FRONT END 4"

ADD HOOD POWER DOME TO
CLEAR & DOUBLE CHOKE
WEBER CARBS

ADDITIONAL SCOOP IN
MODESTY PANEL

LUCAS " FLAME-THROWER
DRIVING LIGHTS
IN CENTER OF GRILL

TOP QUOTE
"The name is often the toughest part of a car to get right. It's easier to design doors and roofs." —Lee Iacocca

FUN FACT
Early on, "Torino" was considered as a name for Ford's new sporty car. As the Italian spelling for the city of Turin, it made a connection to European sports cars. However, the name was rejected when it was learned that Henry Ford II was having an affair with a jet-setting Italian divorcee.

Throughout its development, Ford's new sporty car was known by several names, including the in-house "T-5" code and the more informal "Special Falcon." By late 1963, just a few months before introduction, the need for a name became critical as marketing ramped up its efforts toward promotion and advertising.

There were many suggestions. Henry Ford II offered "Thunderbird II," a designation that was ignored. Dave Ash and Joe Oros pushed for "Cougar," the name for the Ford Studio's winning clay design. They even sent Iacocca a die-cast Cougar grille emblem with the note, "Don't name it anything but Cougar!"

Finally, John Conley from Ford advertising agency J Walter Thompson was dispatched to the Detroit Public Library to compile a list of animal names. From a list that included everything from Aardvark to Zebra, one stood out: Mustang. J Walter Thompson preferred it because it had "the excitement of the wide-open spaces and was American as all hell."

There was also a connection to a Ford two-seater sports car concept that was making the show-car rounds. It was called Mustang, initially suggested by stylist John Najjar to honor the P-51 Mustang fighter planes from World War II. As Najjar related to historian Bob Fria, "My boss, Bob McGuire, thought it was too 'airplaney' and rejected that idea. I again suggested the Mustang name but with a horse association because it sounded more romantic. He agreed and together we selected the name."

The horse name also provided the imagery for the car's grille and emblems, a running pony as penned by stylist Phil Clark for the Mustang two-seater concept, which became known as the Mustang I. Clark's design, which included red, white, and blue bars behind a galloping horse, would be adopted for the new Mustang production car.

THE LAUNCH
MARKETING AND ADVERTISING

With continuing market research indicating that the Fairlane Committee's hunch about a new sporty car was accurate, Ford ramped up one of the most expensive vehicle launches in American auto history. Approaching the Mustang's April 17, 1964, on-sale date, Ford invited magazine writers to Dearborn to emphasize the emerging Baby Boomer market and brought in two hundred of the nation's top radio disc jockeys for a preview drive in the new Mustang. On Thursday night, April 16, Ford bought simultaneous commercial slots for all prime-time TV programming between 9:30 and 10:00 p.m. Plans were also underway for a major press conference at the New York World's Fair and, the following month, to showcase the Mustang as the pace car for the Indianapolis 500.

On April 17, 2,600 major newspapers carried full-page Mustang advertisements. Twenty-four national magazines hit the newsstands with full-page or double-truck spreads showing what Iacocca called the "Mona Lisa look"—a profile of a white Mustang hardtop with minimal copy, just a simple "The Unexpected." Small-car owners around the country found Mustang advertising flyers in their mailboxes. Mustangs were also displayed in two hundred Holiday Inn lobbies and at fifteen of the nation's top airports.

In New York City, Ford went to great heights for a photo opportunity at the top of the Empire State Building. To make it happen, a Mustang convertible was disassembled so the pieces would fit into the elevators, then reassembled on the 102nd floor observation deck.

In a remarkable and unprecedented coup, Ford's public relations department scored simultaneous covers on *Newsweek* and *Time* magazines, both depicting Lee Iacocca with a red Mustang. "I'm convinced that alone led to the sale of an extra hundred thousand cars," said Iacocca.

THE LAUNCH
WORLD'S FAIR INTRO

Way back in 1961, the Fairlane Committee had targeted the Mustang's introduction for the opening of the 1964 New York World's Fair. With the international press gathered in Flushing Meadow, New York, for the fanfare leading up to the Fair's opening on April 22, Ford scheduled a press conference at the Ford Pavilion to introduce the 1965 Mustang on April 13, four days before the car officially went on sale at Ford dealers. Henry Ford II, Lee Iacocca, Don Frey, and other Ford executives were there to describe the slowly spinning Mustang on the stage.

Cleverly, the spring introduction also positioned the Mustang's debut away from the other 1965 new-car introductions, which would happen as usual in the fall. In this case, Mustang got all the attention.

One of the New York World's Fair highlights was the Magic Skyway at the Ford Pavilion. Designed by Disney, the ride transported visitors through a timeline from prehistoric dinosaurs and Stone Age cavemen to life in the future with animated mechanical characters that would become a Disney trademark. The passengers rode in 146 Ford convertibles, including twelve new Mustangs, on a pair of separate moving tracks that ran along glass tunnels around each side of the Pavilion—for maximum visibility to the long lines of fair goers waiting below—before entering the attraction.

When the Mustang assembly line started up on March 9, twelve of the first fourteen completed cars—VINs 100003 to 100014—were convertibles destined for the Magic Skyway. They were delivered to Caron & Company for Skyway preparation, which included the installation of brackets to attach the cars to the tracks, deactivating the brakes and disconnecting the steering linkage, and installing the four-track tape player in the trunk that provided ride narration through the radio speaker

THE LAUNCH
SALES FRENZY

BY THE NUMBERS

8 to 1: Number of 1964 ½–65 Mustangs sold compared to 1964 ½–65 Barracudas, introduced by Plymouth just two weeks before the Mustang. The Barracuda was little more than a fastback roof grafted onto the stodgy Valiant.

FUN FACT

Two months before the Mustang's introduction, the Beatles arrived in America, reaffirming and bolstering a new youth movement as Baby Boomers began influencing everything from music to automobile design.

HISTORIC TIDBIT

Some of the earliest production Mustangs were shipped to the farthest-reaching Canadian coasts so they would be in showrooms on introduction day. Airline pilot Captain Stanley Tucker stopped at George Parsons Ford in St. Johns, Newfoundland, and drove away in a white convertible not knowing that he had purchased VIN 100001, the first serialized Mustang.

After weeks of hiding new Mustangs in storage buildings and even in salesmen's home garages, Ford dealers pulled off the wraps on April 17, 1964, to give the public its first up-close look at Ford's new sporty compact. Thanks to the massive promotion, near-pandemonium broke out, almost like Beatlemania two months earlier. In Chicago, a dealer locked his doors because he feared the mob was endangering both customers and salespeople. In Pittsburgh, a dealer was unable to lower a Mustang from a wash rack because so many people were crowded underneath. In Texas, fifteen buyers got into a bidding war over a dealer's last available Mustang, then the winning bidder reportedly slept in the car overnight as he waited for his check to clear the bank.

Most dealers sold their allotment of Mustangs that first weekend. In one day, Ford salesmen wrote twenty-two thousand Mustang orders, creating a two-month backlog even though Iacocca had had the forethought to add two more assembly plants—San Jose and New Jersey in addition to Dearborn—in an attempt to keep up with the anticipated demand. By the end of Ford's 1964 production cycle in mid-August, Mustang sales had reached 121,538. By the end of the extended seventeen-month 1965 model year, Ford had sold 680,989 Mustangs. A few months after the little-changed 1966 Mustangs arrived, sales topped one million, a record accomplishment for a new car.

While advertising promoted the low $2,300 price, that was for a base six-cylinder hardtop with three-speed stick. Many buyers, however, drove away in Mustangs averaging $1,000 more with high-profit options like 289 V-8, automatic, console, power steering and brakes, styled steel wheels, and vinyl top. For the Mustang's first anniversary, Ford packaged several special or optional components into the GT Equipment and Décor Interior Groups.

Iacocca's plan had worked to the tune of $1.1 billion in net Mustang profits for Ford.

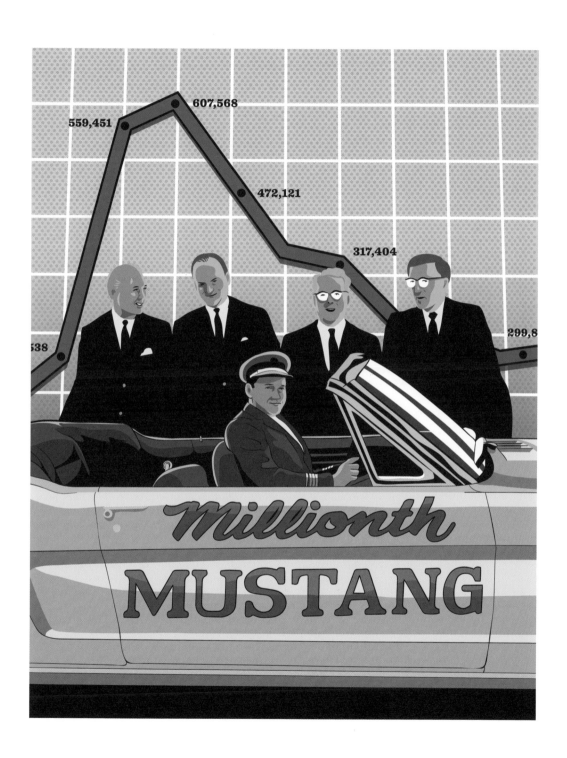

THE LAUNCH
INDY PACE CAR

FUN FACT

The 1964 Indianapolis 500 was won by driver A. J. Foyt, who took delivery of a new Mustang as part of his prize package. He reportedly gave it to his maid.

HISTORICAL TIDBIT

After its pace car duty, the Mustang convertible that paced the 1964 Indy 500 was returned to Holman Moody for servicing, then shipped to Sebring International Raceway, where it served as a parade vehicle and loaner vehicle. It survives today in private hands.

BY THE NUMBERS

1964 ½: Like all early 1965 Mustangs built between March and mid-August 1964, the pace cars became known as 1964 ½ models because they were assembled during Ford's 1964 production cycle. The major difference is 1964's generator charging system, which was replaced by an alternator for 1965.

With its selection as the pace car for the 1964 Indianapolis 500, the Mustang was front and center for the 48th running of America's Great Race on May 30, 1964. Three early production Mustang convertibles were shipped to Ford's NASCAR racing shop Holman-Moody, where the 260 engines were replaced by 289 High Performance versions so the cars could reach the mandated 140 miles-per-hour pace speed. The suspensions were also modified with mismatched shocks for stability in the Indy turns. Reportedly, only two of the cars were completed and delivered to Indy, both Ford Fleet White with Indy Pace Car lettering and blue Rally stripes. VIN 100241 actually paced the race—driven by Henry Ford's grandson, Benson Ford—where it was seen by an estimated three hundred thousand Indy spectators.

For maximum exposure, another thirty-five convertibles were provided for use as VIP cars and parade-lap duty for the Indy 500 festival queens.

Continuing to take advantage of the Indy 500 publicity, Ford produced 190 Mustang Indy Pace Car replica hardtops for a pair of Checkered Flag and Green Flag dealer sales promotions, with the top five performing dealerships in each sales district receiving a car, each painted in a special code C Pace Car White with white and blue interior, 260 V-8, and automatic transmission. Indy 500 Pace Car decals and over-the-top Rally stripes were also part of the package. Lee Iacocca personally presented the keys to around a hundred dealers during a special Checkered Flag delivery ceremony in Dearborn.

Japan's surrender in September 1945, which ended World War II, launched a new era of American prosperity. Not since the Roaring Twenties had America experienced such a feeling of well-being as soldiers returned from the European and Pacific theaters. Manufacturers switched from producing tanks and fighter planes to the peaceful production of stylish automobiles and passenger planes, providing good-paying jobs. Congress passed the GI Bill to provide low-interest loans for homes and higher education. Returning war veterans married their high-school sweethearts, found jobs, and bought homes. The American Dream was in reach for millions.

The millions also started families. In 1946, one year after the end of World War II, 3.5 million births were recorded in the US, nearly three-quarters of a million more than 1945. During the post-war years of 1946 to 1964, America saw a boom of 79 million newborns, a group that would become forever defined as the "Baby Boomers."

The first Baby Boomers, born in the late 1940s, came of age in the 1950s. Unlike their parents, who had worked hard through the Great Depression and survived World War II, Baby Boomers experienced a happier childhood. Families bought their first television sets so they could watch *The Lone Ranger* and *I Love Lucy*. Elvis Presley and Jerry Lee Lewis ushered in the rock-and-roll era, which would provide the Baby Boomer soundtrack for the following decade. Companies catered to Baby Boomer fads with hula-hoops and Davy Crockett coonskin caps. Post-war prosperity put a fancy new car in every driveway as individual mobility became a way of life, leading to drive-in movies and restaurants. Times were good as Americans moved into the 1960s.

The first of the Baby Boomers would reach the age of 18 in 1964. Lee Iacocca and his Fairlane Committee saw it coming. Ford would be ready with the Mustang.

Fairlane Committee: The secret group of Ford executives, led by Lee Iacocca, that met at Dearborn's Fairlane Hotel to discuss their ideas for the car that became the Mustang.

Falcon: A Ford economy compact introduced in 1960. The Falcon chassis and engines would be used for the 1965 Mustang.

Magic Skyway: An attraction sponsored by Ford and designed by Disney for the Ford Pavilion at the 1964–65 New York World's Fair. Guests rode in Ford convertibles, including new Mustangs, to travel through time, from the ancient past and into the future.

Monza: A sportier Corvair model introduced in 1961 that caught the attention of the youth market. Named after the Italian race track, the Monza sold well and influenced Ford's decision to create a sportier compact car.

P-51: A single-seat fighter plane used during World War II and also known as the "Mustang." Several versions were built, including the P-51D with supercharged Packard engine for a top speed of 425 miles per hour.

Pace car: Also known as the safety car, the pace car maintained a safe speed while leading—or pacing—the race cars at the start of a race or during caution periods. The Indianapolis 500 pace cars were typically supplied by manufacturers for marketing visibility.

Sports car: When coined in 1919, *sports car* described a small, two-seater vehicle with good handling. With its sporty looks, the four-place Mustang didn't really fit the definition but nonetheless was often described as a "practical sports car."

World's Fair: A universal exposition, held every several years in different parts of the world, to showcase the accomplishments of nations. The first one was held in 1851 in London. In 1964, the New York World's Fair served as the marketing launch for the 1965 Mustang.

GENERATIONS

FIRST GENERATION

Although little more than a new long-hood and short-rear-deck body on the Falcon chassis, the 1965 Mustang was a runaway success, arriving just as the oldest Baby Boomers reached car-buying age. However, the sporty image of the 1965–66 Mustang also appealed to other demographics, even grandparents. In particular, women fell in love with the Mustang's sexy lines, compact size, secure handling, and great economy with its base six-cylinder engine. The available 289 V-8 became synonymous with Mustang performance, especially the high-performance version with 271 horsepower. To celebrate the Mustang's first anniversary in April 1965, Ford introduced a pair of snazzy option packages—the GT Equipment Group and Interior Décor Group.

Typical of the 1960s, the Mustang was restyled for 1967–68, gaining size and heft to accommodate a 390-cubic-inch big-block engine to compete with the arrival of the Chevrolet Camaro, Pontiac Firebird, and even corporate cousin Mercury Cougar. At the same time, designers preserved the Mustang's original pony car character and the interior finally shed its Falcon roots for a fresh all-Mustang cockpit. Updates for 1968 were minor, mainly grille and side-ornamentation revisions.

Change came to the Mustang again for 1969–70. A sheetmetal overhaul delivered a more muscular appearance, especially for the new "SportsRoof" fastback. By the end of 1969, the Mustang was available with eleven engines, from the six-cylinder to the 428 Cobra Jet, including the midyear Boss 302 and Boss 429, both created to legalize engines for racing. Two new models were added to the lineup—a Mach 1 for a "speed of sound" image and luxurious Grande for the hardtop. Both continued into 1970 on a mild Mustang redesign.

The Mustang grew even larger for 1971–73, reaching almost intermediate size to accommodate larger engines such as the 429 Cobra Jet. However, with ever-increasing government emissions regulations, the 1971 Mustang was the last available with a big-block; the 351 Cleveland engine became the top performance offering for 1972–73.

GENERATIONS
MUSTANG II

HISTORICAL TIDBIT
Two months after the introduction of the 1974 Mustang II,
American fuel prices soared
during the OPEC oil embargo.
Once again, Iacocca's hunch
was correct—the smaller, more
fuel-efficient Mustang II was
the right car at the right time.

BY THE NUMBERS
385,993: Number of Mustang IIs
sold for 1974, making it the
fourth best-selling Mustang of
all time

FUN FACT
In the popular *Charlie's Angels*
TV show, Farrah Fawcett's
character, Jill Munroe, drove a
white-with-blue Cobra II. The
exposure no doubt led to more
sales for the Mustang II.

The 1965 Mustang was Lee Iacocca's baby, but the Mustang of 1971–73 was not the small, sporty car he had originally envisioned. Iacocca once described the early 1971 Mustang as a "fat pig" and, with the 1969 ouster of Ford president Bunkie Knudsen, who had pushed for the larger size, he was free to investigate a Mustang overhaul that would return the Mustang to its small-car roots. Sagging sales supported his decision.

The 1974 Mustang was such a radical departure from 1971–73 that "II" was added to the name. While retaining its long-hood, short-deck pro-file, the Mustang II was 18 inches shorter and 4 inches narrower than the previous 1973 model. Instead of channeling muscle, Ford promoted the Mustang II as a small luxury car. In its first year, the only available engines were the 85-horsepower four-cylinder and the optional 105-horsepower V-6, establishing the 1974 model as the only Mustang without an available V-8.

Iacocca referred to the Mustang II as his "little jewel," a new kind of American compact with upgraded sound deadening, improved fit and fin-ish, and the first-time use of rack-and-pinion steering and standard front disc brakes. A Ghia model replaced the Grande to add European elegance with pin striping, wire wheelcovers, and a plush interior.

The Mustang II evolved over its five-year production span. Following complaints about the lack of power, a two-barrel 302-cubic-inch V-8 was added to the option list for 1975. In 1976, a Cobra II model with spoilers and Shelby-like stripes projected a performance image, one that evolved into the 1978 King Cobra with yards of colorful striping and a huge snake image on the hood.

While the 1974–78 Mustang II sold well, it was not popular with tra-ditional Mustang enthusiasts. However, Iacocca's "little jewel" kept the Mustang name alive during a tough time in American auto history.

FOX-BODY

FUN FACT
Developed for the Mustang, the Fox-body chassis would eventually underpin other Ford vehicle lines, including the Fairmont/Zephyr, Thunderbird/Cougar, and Lincoln Mark VII.

BY THE NUMBERS
5.0: Liter displacement for the 302-cubic-inch V-8, a high-output engine that would return the Mustang to performance prominence in the 1980s

KEY PERSON
Ford design executive Jack Telnack returned from Ford of Europe in 1976 to take over Ford's light car and truck design. Utilizing his experience with aerodynamics, he influenced the sleek design of the 1979 Mustang.

The Mustang that arrived in Ford dealer showrooms for 1979 was nothing like the Mustangs that came before it. Taking on a more European appearance, the 1979 model lost its traditional pony car cues in favor of a slanted front end, a lower hood, and slab sides for sleeker aerodynamics and resulting fuel economy gains. Even the Mustang scripts and running horse emblems were replaced by a simple Ford oval on the hood. "We had done about as much as we could with those original design cues from 1964," explained one design manager.

Codenamed "FOX" during its development, the resulting all-new Mustang became known as the Fox-body during its fifteen-year tenure. Offered as a hardtop and new hatchback, the Mustang was once again a sales success, with 369,936 sold during its first year, 177,500 more than 1978. When the Fox-body was selected as the pace car for the 1979 Indianapolis 500, Ford produced a special run of Indy Pace Car replicas with an air dam and rear spoiler that previewed future performance Mustangs.

From 1979 to 1981, Fox-body Mustangs were powered by low-output engines, including a troublesome turbocharged four-cylinder and a meek 4.2-liter V-8. A performance revival started in 1982 with the return of the Mustang GT and an optional two-barrel 5.0-liter "High Output" rated at 157 horsepower. The HO evolved over the next five years, progressing to a four-barrel Holley carburetor in 1984, gaining fuel injection in 1986, and maxing out at 225 horsepower for 1987–93.

After nearly a decade of fuel economy and emissions priorities, the Fox-body brought excitement back to Mustang. The convertible returned in 1983, Special Vehicle Operations offered a turbocharged SVO model from 1984–86, and Ford closed out the era with a 1993 Cobra. Good times had returned for Mustang—and it would only get better.

TOP QUOTE

"I'd be honored to try to save the Mustang." —John Coletti, when offered the opportunity to manage a Ford skunk works group to explore options for continuing the Mustang as a rear-wheel-drive vehicle

TECHNOLOGICALLY SPEAKING

The 1999–04 SVT Cobras were the first Mustangs equipped with an independent rear suspension for better handling and improved ride quality.

HISTORICAL TIDBIT

When Mustang enthusiasts complained that the 1994–95 vertical tri-lens taillights weren't true to the originals from the 1960s, Team Mustang made them horizontal for 1996.

The Mustang faced a crisis in the late 1980s: Ford planned to replace the rear-wheel-drive platform with a Mazda front-wheel-drive chassis. Mustang fans rose up in protest, firing off letters to Ford World Headquarters, demanding, "No Mazda Mustang!" Acknowledging the pressure, Ford vice president Alex Trotman approved a skunk works program, called Team Mustang and headed by John Coletti, to explore the possibility of reworking the existing rear-wheel-drive Fox-body into a new Mustang that could meet increasingly stringent government safety regulations. Coletti succeeded, eventually bracing, strengthening, and modifying the older chassis so much that it became known as the Fox-4, codenamed SN-95.

The latest Mustang arrived as a 1994 model in coupe or convertible body styles. Outwardly, there was no resemblance to the previous generation. Instead, Team Mustang brought back the classic Mustang styling cues, including a mouthy grille opening, side sculpturing, and tri-lens taillights. The interior elaborated on the original Mustang's cockpit-style design. Engine offerings were simplified to two—a V-6 for the base model and continuation of the 5.0-liter V-8 for the GT, with the Ford Special Vehicle Team (SVT) adding a 240-horsepower Cobra model later in the model year.

The pushrod 5.0-liter was retired after 1995 as the 1996 Mustang switched to Ford's new "modular" 4.6-liter V-8s, either two-valve for the GT or high-revving four-valve for the SVT Cobra. For 1999, the Mustang's body was updated with sharper lines to reflect Ford's "New Edge" styling, and the GT's 4.6 was updated to three-valve cylinder heads for more power. Supercharging boosted the 2003–04 Cobra to 390 horsepower.

As the SN-95 reached the end of its production cycle, Ford pumped excitement into Mustang with a pair of special models—a 2001 Bullitt GT that paid tribute to Steve McQueen's fastback in the 1968 movie *Bullitt* and a 2003–04 Mach 1 with a Shaker hood throwback to 1969–70.

GENERATIONS
S197

KEY PERSON

Hau Thai-Tang was a small child in South Vietnam when he saw drag-racing Mustangs that were part of a US military tour. He would become Mustang chief engineer in 2002 during the final development of the 2005 Mustang.

TOP QUOTE

"If you don't get it right, you've got eight million Mustang fans to answer to." —J Mays, Ford group vice president of design

HISTORICAL TIDBIT

Production of the 2005 Mustang moved to Ford's new AutoAlliance assembly plant, ending the forty-one-year run at the old Dearborn Assembly Plant where Mustangs had been built since 1964.

After twenty-five years of Fox-body Mustangs, Ford recognized the need to bring the Mustang into the new millennium on a stronger chassis and modern coil-over MacPherson strut suspension. The perfect starting point was found in the new and modern DEW98 platform developed for the Lincoln LS. By the time Team Mustang finished adapting the DEW98 for the Mustang, it had changed so much that it earned its own S197 codename.

For what would become the 2005 Mustang, the designers were challenged to continue the retro styling cues that had proved so popular from 1994–04 while modernizing at the same time. The result was a completely overhauled body based loosely on the muscular 1967–68 Mustang and incorporating a mouthy grille opening, outboard headlights, side sculpturing, and tri-lens taillights. The 2005 Mustang debuted as a coupe with the convertible added later in the model year. A newer V-6 powered the base model, while the GT got an upgraded 4.6-liter V-8 with 300 horsepower.

SVT waited two years to reveal its high-performance S197 Mustang, renaming the Cobra as a 2007 Shelby GT500 with a 500-horsepower supercharged 5.4-liter V-8.

A minor styling update for 2010 provided the Mustang with a leaner and meaner demeanor. The following year, the 5.0-liter name reappeared as a new four-valve "Coyote" V-8 with 412 horsepower. The 302-cubic-inch displacement opened the door for a new Boss 302, produced in 2012–13 as a track-ready Mustang. SVT stepped up its game with the 2013–14 Shelby GT500, powered by a supercharged 5.8-liter V-8 with 662 horsepower, a new high-water mark for Mustang output. Ford also brought back two names from Mustang's past—the California Special in 2007 and a reprise of the Bullitt GT for 2008–09.

GENERATIONS
S550

"Going Global" was the mantra behind the sixth-generation Mustang, destined to debut as a 2015 model in time for the Mustang's 50th anniversary. Once again, Ford gambled with a major overhaul and redesign for an iconic car that was coveted not only in the United States but also around the world. Moving the Mustang forward meant retaining the pony car flavor and heritage while at the same time improving performance, handling, comfort, safety, and fuel economy. There were two edicts from upper management: replace the rough-riding solid rear axle with a modern independent rear suspension and develop available right-hand drive to appeal to buyers in countries such as Great Britain and Australia.

Adding IRS to the 2015 Mustang, codenamed S550 during its development, led to a completely new front suspension as well. The stylists succeeded in maintaining the Mustang character while pushing in a more modern direction to appeal to younger buyers. Ford's corporate oval grille opening left little room for Mustang interpretation, but the coupe's sloping roofline and slanted rear panel with tri-lens taillights closely resembled the 1969 fastback. The convertible received its own upgrades, including a cloth top, quicker up-and-down operation, and single-handle latch.

As before, the base Mustang came with a V-6 and the GT was powered by the V-8 (in this case a 435-horsepower Coyote 5.0-liter), but Ford also added a third powertrain in the form of an EcoBoost 2.3-liter. The turbocharged four-cylinder combination produced 310 horsepower and 31 miles-per-gallon fuel economy.

Making a big splash during the Mustang 50th anniversary was a 50 Year Limited Edition, offered to only 1,964 buyers in Wimbledon White or Kona Blue with vintage design cues. For 2016–18, Ford SVT produced a Shelby GT350 and GT350R as the next-generation Mustang track cars with high-tech MagneRide dampers and a 5.2-liter V-8 with a flat-plane crankshaft as used in racing exotics.

THE SAME, ONLY DIFFERENT

Rarely do automotive nameplates survive for more than a decade in their original configuration. Ford's Fairlane spent its formative years as a full-size car before morphing into a mid-size for 1963. Mercury's Comet started out as a Falcon-based compact; by 1966, it was also an inter-mediate. Amazingly, the Mustang has survived for over fifty years as a sporty two-door while adapting to tightening emissions and safety requirements, market trends, and one fuel crisis after another.

In its original form, the Mustang was basically a rebodied Falcon. Yet unlike the Falcon, which was primarily a low-priced economy car when it debuted in 1960, the Mustang appealed to a wider group of buyers. The long-hood, short-rear-deck profile added a sporting image, while a long list of options allowed buyers to customize their Mustang for econo-my, luxury, performance, or even racing. Unlike the Corvette, the Mustang came with a rear seat and a trunk, making it practical for double dating or a trip to the grocery store. Mustangs were purchased as a second car for the family.

Interestingly, the Mustang has adapted to the times. The larger 1971–73 model provided room for large performance engines, the smaller Mustang II arrived just in time for higher fuel pric-es, the inexpensive and lightweight 5.0-liter

Fox-body played a huge role in the 1980s performance revival, and the retro-styled S197 appealed to the Baby Boomers who loved the Mustang in the 1960s. Although larger than the original 1964 Đ, today's Mustang remains true to the original mission as a two-door, four-seat sporty car that can be many things to many people, including those who live in other countries.

As noted by *Car & Driver* magazine, "Few things have held up over the past fifty years as well as the Mustang."

GLOSSARY: GENERATIONS

EcoBoost: Ford's marketing name for its line of small-displacement, turbocharged engines. A 2.3-liter EcoBoost four-cylinder was added as a new model for the Mustang during 2015 to 2018. It developed 310 horsepower with over 30 mpg fuel economy.

Fox-body: Derived from the codename FOX as used by Ford to describe the 1979 Mustang chassis. It would also be used for other Fords, including the Granada and Thunderbird.

Ghia: The 1974 to 1978 Mustang II replacement for the Grande, named for the Italian styling studio that created one of the first Mustang II designs.

Grande: A luxury model for the Mustang hardtop, offered from 1969 to 1973 with décor interior, pin stripes, extra insulation, and other extras.

GT Equipment Group: A package for 1965 to 1969 Mustangs that included a group of visual and functional options and other components, which varied from year-to-year. It typically added the heavy-duty suspension, side stripes, and fog lamps.

Independent rear suspension: Rear wheels are independently sprung, unlike the solid axle arrangement found in Mustangs prior to 2015. The IRS for 2015 to 2018 Mustangs provides improved handling and a more comfortable ride.

Interior Décor Group: An upgraded interior package for the first generation Mustangs, typically adding woodgrain trim, molded door panels, and other upgrades depending on the model year.

S197: The fourth generation Mustang chassis, introduced as a 2005 model and continuing through 2014.

S550: The codename for the fifth generation 2015 Mustang, updated with a completely new chassis, suspension, and retro body.

SN-95: Ford's in-house code name for the Fox-body update that would become the 1994 to 2004 Mustang.

SportsRoof: Ford's nomenclature for the fastback body style from 1969 to 1973.

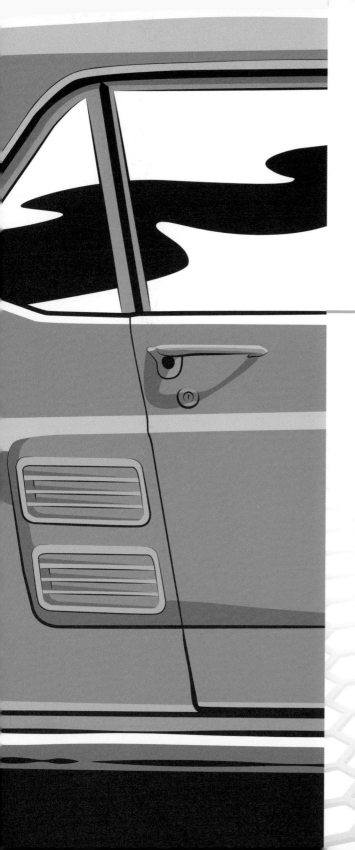

EVERY MAN'S SPORTS CAR

EVERY MAN'S SPORTS CAR
HARDTOP

Lee Iacocca called it the "Mona Lisa look." A profile photo of a white hardtop, used extensively in early advertising, dramatized the 1965 Mustang's long-hood ("Implies there's a lot of engine under there," said one Ford executive) and short-rear-deck styling with a formal roofline (very similar in shape to the earlier Lincoln Continental), a look that implied high style and luxury. In truth, most 1965–66 Mustang hardtops were delivered as low-cost base models with a six-cylinder engine and three-speed manual transmission. Without them, Mustang would not have topped 1 million sales in less than two years. Nearly 70 percent of 1965–73 Mustangs were hardtops, also known as "coupes" or "notchbacks."

During 1969–73, the hardtop earned its luxury reputation by becoming the foundation for the Grande, a model that packaged a soft-riding suspension, extra insulation, and Deluxe interior with woodgrain trim, cloth high-back bucket seats, and a console. During the 1974–78 Mustang II years, the top-of-the-line Mustang was the hardtop Ghia, named after an Italian design company and the epitome of Iacocca's vision for his smaller Mustang—European elegance with a half vinyl roof, opera window, and crushed velour upholstery.

Although Ford abandoned traditional Mustang styling cues during the 1979–93 Fox-body era, the hardtop soldiered on as the base model. Most buyers drove away from Ford dealerships in low-priced base models with four-cylinder power, while the Ghia continued to offer European-type luxury. An available carriage roof added a convertible-top look. Starting in 1986, performance enthusiasts knew to choose the LX 5.0-liter hardtop for its low cost and lighter weight. For police pursuit duty, a Special Service Package hardtop supplied a 5.0-liter engine, fluid coolers, single-key locks, and unique equipment such as silicone hoses and a heavy-duty alternator.

With the Mustang's 1994 redesign, the hardtop became more of a coupe when the formal shape was replaced by a retro fastback look.

EVERY MAN'S SPORTS CAR
CONVERTIBLE

For many in 1964, the lasting image of the Mustang was a white convertible driven in the Swiss Alps by a beautiful blonde. The James Bond film *Goldfinger*, released in September 1964, featured the Mustang in a scene with actress Tania Mallet's convertible dueling with Sean Connery's Aston Martin, a chase that ended when 007's wheel spinners popped out to shred the Mustang's rear tire.

In the 1960s, nearly every car line offered a convertible, even Ford's economy-based Falcon. The Mustang's sporting demeanor was ideal for top-down fun, and the hardtop's formal roofline lent itself well to an attractive convertible model with the top up. For the first four months of 1964 production, the 1965 Mustang was available only as a hardtop or convertible, which had manual up/down operation, power assist optional. By 1973, the Mustang was the only Ford available as a convertible.

The convertible disappeared during the 1974–78 Mustang II era; open-top driving was reduced to a sunroof or, starting in 1977, a T-top hatchback. As the Mustang emerged from ten years of emission and safety priorities, Ford took extreme measures to bring back the convertible for 1983 by sending roofless hardtops to nearby Cars and Concepts for conversion into GT or GLX convertibles. At least the convertible was back.

The 1994 redesign for the Fox-body was particularly good for the Mustang convertible. The new wedge-shaped body provided a sleek top-down appearance, enhanced by a top that stacked deeper into the body. Mass dampener technology was also employed to cancel out the cowl shake that had plagued the 1983–93 convertibles.

Delayed several months, the arrival of the 2005 Mustang convertible, based on the new S197 platform, was treated like a separate launch and lauded as the best top-down model ever, with a stiffer chassis and less wind buffeting at speed. The all-new 2015 convertible was upgraded with a cloth top, quicker up-and-down operation, and a single-lever latching system.

EVERY MAN'S SPORTS CAR
FASTBACK

Initially, Ford planned to offer the 1965 Mustang in two body styles, hardtop and convertible. But when designer Gale Halderman, who had sketched the original Mustang shape, proposed a fastback in May 1962, the sloping roof was added to the Mustang lineup. Introduced in September 1965, five months after the hardtop and convertible, the "2+2" featured functional air-extracting C-pillar louvers and fold-down rear seats that opened into the trunk for stowing golf or skis. The fastback was perfect for Ford's Total Performance marketing campaign, especially when equipped with the 289 High Performance (Hi-Po) or GT Equipment Group. Hi-Po fastbacks were also the starting point for Shelby American's GT350s.

With the 1967–68 redesign, the fastback's roofline extended from the top of the windshield to the rear panel. Later, Ford marketed the more muscular 1969–70 fastback as the SportsRoof, with its roofline ending at the rear panel with a built-in ducktail spoiler. The SportsRoof was the body style for Mustang supercar models, including the Mach 1, Boss 302, and Boss 429. Popular SportsRoof options included a trunk-mounted rear spoiler and Sport Slats, also known as rear window louvers, that covered the large rear glass to provide some relief from the sun.

The SportsRoof continued for the larger 1971–73 Mustang, becoming more of a "flatback" with a huge rear window that was angled so much that some described the view from the interior as a "mail slot."

The fastback was replaced by the hatchback during the Mustang II and early Fox-body generations. Starting with the 1994 SN-95's hybrid hardtop/fastback roofline, "coupe" became Ford's preferred description instead of fastback. However, the fastback name returned for the 2015 Mustang to describe the roofline that closely resembled the 1969–70 SportsRoofs.

EVERY MAN'S SPORTS CAR
OPTIONS AND ACCESSORIES

Long before it had a name, Ford's sporty compact had a goal as established by Lee Iacocca's Fairlane Committee in late 1961: "one basic car with many available options." "Designed to be designed by you" was the marketing theme to describe the first American car offered as a base model—with standard items such as bucket seats and carpet—but with a long list of available extra-cost equipment. Radio announcers read from Ford's advertising copy: "You can tailor your Mustang to your own personal needs and tastes with a choice of options previously unprecedented in a car with this low price."

At its introduction on April 17, 1964, the base Mustang came with a six-cylinder engine and three-speed manual transmission. From there, buyers could choose from numerous options, including V-8 engines, automatic or four-speed transmissions, an AM/FM or AM/eight-track radio, various wheel covers, and air conditioning, plus nearly fifty dealer-installed accessories such as a rear seat speaker, sun visor vanity mirror, and remote control trunk release, even a tissue dispenser. The number and variety of body styles, exterior color choices, options, and accessories allowed buyers to make their Mustang their own.

Ford simplified the option selection during the late 1980s by curtailing separate options and grouping many together as Preferred Equipment Packages (PEP). For example, PEP 249A for the GT added air conditioning, Power Equipment Group (windows, door locks, and so on), speed control, and an AM/FM/cassette stereo with Premium Sound. During the S197 era, Mustangs were available in Standard, Deluxe, and Premium models, each with its own package of upgrades. For the 2015 and later Mustangs, EcoBoost and GT models were available in Premium trim that added heated and cooled leather seats and a nine-speaker stereo. An available 401A Equipment Group upgraded to Shaker Pro Audio and other features. High-tech options such as Electronic Line-Lock and Selectable Drive Modes were unheard of in 1964.

EVERY MAN'S SPORTS CAR
ECONOMY

In its most basic utilitarian form, and true to its Falcon heritage, the Mustang was an economy compact with a sports car body. Early advertising promoted the $2,368 suggested retail price for a standard hardtop with a 170-cubic-inch six-cylinder, three-speed manual transmission, and black sidewall tires. And although the 170 was replaced by a 200-cubic-inch version for 1966, the inline six-cylinder offered good fuel economy combined with peppy performance. It would serve as the Mustang's base engine until 1971, when it was replaced by a larger 250-cubic-inch inline-six to power the larger and heavier 1971–73 Mustangs.

Ford's "right car at the right time" slogan would prove perfect for the 1974 Mustang II. When the OPEC oil embargo in the fall of 1973 sent fuel prices soaring, the smaller, lighter Mustang II was already in showrooms with a standard 2.3-liter four-cylinder or available 2.8-liter V-6. And although the two-barrel 302-cubic-inch V-8 returned in 1975, the Mustang II's fuel efficiency was the primary reason that Ford sold over 1.1 million during its five-year production span.

By the late 1970s, it was obvious that fuel prices would never return to less than 30 cents per gallon as in the 1960s. Ford responded with the 1979 Mustang's slanted front end and lower hood line for improved aerodynamics and resulting fuel savings at speed. Even as a new 5.0-liter High Output signaled a return to performance in 1982, Ford was bragging about the 2.3-liter four-cylinder's 32 miles per gallon.

Performance enthusiasts were concerned about the switch to fuel injection for 1986, but it would prove to be the best thing that could have happened for both Mustang horsepower and fuel economy. By 2013, the supercharged Shelby GT500 was tuned for 662 horsepower and 24 miles per gallon on the highway, good enough to avoid the government's dreaded gas-guzzler tax.

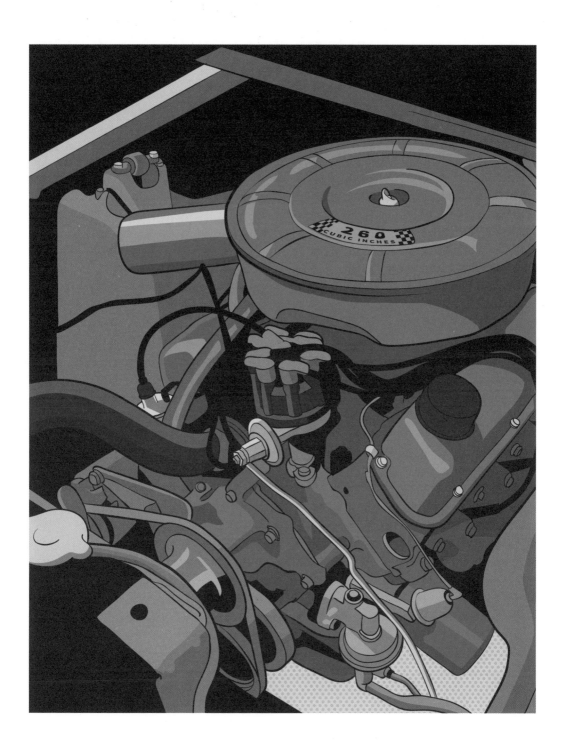

EVERY MAN'S SPORTS CAR
HATCHBACK

FUN FACT
Jim Wangers, the automotive marketing executive who created the Pontiac GTO in 1964, was involved in the creation of the hatchback-based 1976 Cobra II.

HISTORICAL TIDBIT
When the 1979 Mustang was selected as the pace car for the Indianapolis 500, Ford produced 10,487 Pace Car hatchbacks in Pewter Metallic with orange, red, and black graphics.

TECHNOLOGICALLY SPEAKING
In 1986, new government safety standards mandated a third brake light. For Mustang hatchbacks, it was mounted in the rear spoiler.

A new kind of fastback debuted for the 1974 Mustang II. Called "3-Door 2+2" by Ford, the latest hatchback body style took advantage of the sloping roofline by adding a large swing-up rear door for much-improved access to the interior. The Mustang's first-time use of hydraulic lifts held the big door open for loading and unloading. There was no enclosed trunk, but with the rear seat back folded down, the Mustang II hatchback offered 27 cubic feet of storage space, enough for "golf clubs or scuba gear," according to the sales brochure. The hatchback was also the body style of choice for Mustang II performance image models, including the Mach 1, 1976–78 Cobra II, and 1978 King Cobra.

The hatchback continued into the Fox-body era, starting as a 1979 "Three Door" (also called "Third Door Liftgate") for 32.4 cubic feet of storage with the rear seat back folded down. The 1979 Cobra was based on the hatchback, while the 1979 Pace Car's front air dam and rear spoiler previewed the race car look that would appear on the 1982 GT. An aero nose, lower side scoops, and louvered taillights gave the GT hatchback a totally new appearance from 1987 to 1993. The high-tech 1984–86 SVO was also based on the hatchback, one with the Mustang's only use of a dual-plane rear spoiler.

With the Mustang's major redesign for 1994, the hatchback was discontinued, having served its practical purpose for the twenty years spanning 1974 to 1993.

EVERY MAN'S SPORTS CAR
MODERN COUPE

After nearly a decade and a half, the Mustang hatchback was replaced in 1994 by a more traditional coupe as part of the major SN-95 overhaul. But it was not the formal-shaped roof like previous Mustangs. With Mustang body style choices reduced from three to two—coupe and convertible—the new coupe roofline sloped downward toward the trunk with a C pillar and stationary side window that mimicked the 1967–68 Mustang fastbacks. It was not a true hardtop nor a true fastback. The new roof sat on the SN-95's wedge-shaped body and was not integrated into the rear quarter panels.

The SN-95 coupe roof looked very much like a convertible's removable hardtop. In fact, Ford offered a removable hardtop for the 1995 SVT Cobra convertible. With the top installed, the convertible looked just like a hardtop. It was an expensive option; as a result, only 499 were sold.

For the 2005 Mustang coupe, Ford kept the hybrid hardtop/fastback look with a slightly more sloped roofline and C pillars that once again flowed into the rear quarter panels as part of the overall stronger body structure. Addressing one of the Mustang's longtime negatives, the S197 coupe roof added ½ inch to the front seat headroom. A body update for 2010 incorporated Ford's New Edge styling, but the roof would remain the same throughout the S197's fourteen-year production span.

From the rear, the new 2015 Mustang coupe recalled the fastbacks of the 1960s with a sleeker roofline that merged into an angled rear panel with tri-lens taillights. In fact, Ford dropped the coupe designation to bring back the fastback description for the first time since 1968.

SOMETHING FOR EVERYONE

Throughout its fifty-year-plus history, the Mustang has remained true to its "one basic car with many available options" goal, allowing the original pony car to appeal to many people for many purposes, from driving to the grocery store to cruising the high banks at Daytona.

Starting with a base model and adding options as needed (or wanted) was a new marketing concept in 1964. Unlike most cars of the era, especially the budget-priced compacts, the Mustang came standard with amenities such as carpet and vinyl seats, and buyers could choose from many options and dealer accessories. Packaged equipment options (GT Equipment, Décor Interior, and Convenience Groups, for example) became popular over time, eventually leading to Preferred Equipment Packages and the latest Premium models. Today's base Mustang comes with plenty of extras, including power windows and air conditioning, with the Premium model adding leather seats and high-tech electronics.

Interestingly, the most popular Mustang options over the years can be placed into three categories—convenience, luxury, and performance.

Early on, Mustang convenience meant options such as cable-adjusted side mirrors or a remote trunk-lid release. The arrival of power windows in 1971 ushered in switch-controlled conveniences such as power seats, door locks, and mirrors, with all becoming common during the Fox-body era and standard equipment for later models.

Mustang luxury included both appearance and function. Fake woodgrain implied high style, and seat upholstery progressed from the Mustang II's crushed velour to real leather. Air conditioning was available from the beginning, albeit as a Falcon-like add-on that hung under the dash, but quickly evolving into in-dash units. Newer Mustang air conditioning even comes with automatic climate control.

For Mustang, performance options started with the 289 High Performance engine, four-speed transmission, and Rally-Pac tachometer, then progressed steadily through 428 and 429 Cobra Jets before the 5.0-liter HO became the standard bearer for the Fox-body GT and LX 5.0. For 2018, the GT is powered by a 460-horsepower Coyote 5.0-liter with available ten-speed automatic transmission.

GLOSSARY: EVERY MAN'S SPORTS CAR

Carriage roof: A type of automotive roof styling designed to give the appearance of a convertible. It was available for the 1980 Mustang hardtop.

Cars and Concepts: An automotive subcontractor that converted 1983 Mustang hardtops into convertibles, making them the first top-down Mustangs since 1973.

Convertible: A body style with a top that folds back or retracts to provide open cockpit driving.

Ducktail spoiler: As found on the 1969–70 Mustang SportsRoofs, a type of built-in rear air-foil with an upswept design to add more aerodynamic downforce to the rear of the car at speed.

Fastback: A roof style that slopes toward the rear of the car to give an appearance of sleek aerodynamics.

Hardtop: The most popular type of roof for American automobiles, typically found on two- and four-door vehicles and also described as coupe or notchback. It was usually the base body style for the Mustang.

Hatchback: A body style with a "third door" that opened from the rear for access to the interior's rear storage area.

Light bar: An aftermarket bar created for Mustang convertibles by Classic Design Concepts. Not designed for structural or safety purposes, it was more for aesthetics and described as a *light bar* due to its third brake light.

OPEC: The Organization for Petroleum Exporting Countries, which initiated an oil embargo in October 1973 right after the introduction of the smaller and more fuel-efficient 1974 Mustang II.

Opera window: Small porthole-size windows installed in vehicle C-pillars to imply luxury. They were popular in the 1970s and used for Ghia hardtops during the Mustang II era.

T-Top: A type of roof with a removable panel on each side. Offered for Mustangs during the Mustang II and Fox-body eras.

Velour: A fabric with a pile or napped surface to resemble velvet, commonly used for automotive upholstery in the 1970s for a luxurious feel and look.

PERFORMANCE

PERFORMANCE
TOTAL PERFORMANCE

Since mid-1957, Ford and the other American auto manufacturers had abided—some more loosely than others—by an Automobile Manufacturers Association resolution that prevented the companies from participating in racing, including the publicizing of race results and advertising speed-related cars. The manufacturers loathed the ban. NASCAR in particular was popular among the general public, and the Big Three viewed racing as an opportunity to increase sales with a "Win on Sunday, Sell on Monday" mindset. Ironically, Ford chairman and CEO Henry Ford II was serving as president of the AMA when, in 1962, he advised the organization, "The resolution adopted in the past no longer has either purpose or effect. Accordingly, we are withdrawing from it."

The Mustang was still on the drawing board when Ford pulled away from the AMA racing ban. Free to unleash its engineering and marketing forces on a public that was craving speed for both street and track, Ford realized that its new sporty car could play a huge role in a new marketing campaign. In the twelve months prior to the Mustang's introduction, Ford transitioned its corporate advertising from "Lively Ones" to "Total Performance." When the Mustang arrived on April 17, 1964, new models were often displayed in dealer showrooms adjacent to 427 Galaxies and 289 Hi-Po Fairlanes. Two months later, Ford GT-40s appeared at Le Mans for the first time. By the end of the year, Ford Galaxies had racked up thirty wins to secure the NASCAR manufacturer's championship once again.

By the mid-1960s, the Mustang was smack-dab in the middle of Ford's Total Performance campaign. Carroll Shelby's GT350 won the SCCA's B-Production championship in 1965, defeating Corvettes in the process, and by 1966, driver Jerry Titus was sitting atop his Shelby-prepared hardtop with the Trans-Am championship trophy. It was only the beginning for a Mustang performance tradition that would drive sales for over fifty years.

PERFORMANCE
289 HIGH PERFORMANCE

During the Mustang's first two months, the top-rated engine was a 210-horsepower 289. In June 1964, Ford injected more muscle by adding the Fairlane's 289 High Performance to the option list.

Rated at 271 horsepower at 6,000 rpm, the 289 Hi-Po, as it was known, was prepped for high revs with a solid-lifter camshaft, a dual-point distributor, a 595-cfm four-barrel carburetor, header-style exhaust manifolds, and cylinder heads with smaller combustion chambers, cast spring cups, and screw-in rocker arm studs. Inside, the Hi-Po short-block sustained the high-rpm power with thicker main bearing caps, a high-nodularity crankshaft, and larger rod bolts, which mandated revised crankshaft counterweighting and a new balancer. Hi-Po 289s were also equipped with a high-revving water pump and larger pulley for the generator or alternator.

Available for all 1965–67 Mustangs, the 289 High Performance was a package deal, not only adding the 271-horse small-block but also supplying a mandatory four-speed, 9-inch rear end, dual exhaust (an Arvinode system from October 1964 to March 1965), and a heavy-duty suspension with stiffer springs and shocks, a larger front sway bar, quicker ratio steering, and 6.95×14-inch tires. Peering under the hood, the Hi-Po was identified by its chrome valve covers and open-element air cleaner. Due to the high-rpm capabilities, air conditioning and power steering were not available. A small "High Performance" metal plate with checkered flags, sandwiched behind the standard 289 emblem and the front fender sheet metal, was all that identified the 289 High Performance Mustang.

In 1966, Ford beefed up the C4 Cruise-O-Matic transmission for use behind the Hi-Po.

Although high in both performance and marketing visibility, only 13,231 1965–67 Mustangs were factory equipped with the 289 High Performance engine—7,273 for 1965, 5,469 for 1966, and just 489 for 1967, a year when the high-revving Hi-Po was overshadowed by the high-torque 390 big-block.

PERFORMANCE
COBRA JET

Adding a 320-horsepower 390 to the Mustang's powertrain list for 1967 was certainly a step in the right direction in a time when muscle cars were growing in popularity. However, the 390 was basically a high-torque passenger car engine with limited horsepower capabilities due to its restrictive cylinder heads. In *Hot Rod* magazine, influential Ford dealer Bob Tasca criticized the 390's lack of performance potential in an article that reached the top floors of Ford World Headquarters. The result was a four-year corporate plan for Ford performance, including new image models, increased capabilities for producing parts, an energized marketing and advertising campaign, and the availability of Cobra Jet engines.

The 428 Cobra Jet arrived in April 1968 as an option for Mustangs equipped with the GT Equipment Group. Based on the existing 428 Police Interceptor, the CJ made more usable power with updated 427 Low-Riser heads, a 390 GT camshaft, a cast-iron intake with a 735-cfm Holley four-barrel, and low-restriction exhaust manifolds. It was rated at a conservative 335 horsepower, a sly deception to avoid scrutiny from insurance companies while also gaining an advantage in organized drag racing.

The FE-based 428 CJ continued into 1969–70 as an option for all Mustangs, either Q-code or R-code with Ram-Air, before being replaced in 1971 by the 429 Cobra Jet based on Ford's new 385-series big-block. With canted-valve heads, the 429 CJ produced 370 horsepower, which was upped to 375 as a Super Cobra Jet with a solid-lifter camshaft and Holley four-barrel as part of the Drag Pack option.

The 1971 429 Cobra Jet was the last big-block engine offered for the Mustang. For 1972–73, the Cobra Jet name shifted to the four-barrel 351 Cleveland.

PERFORMANCE
MACH 1

Mustang was moving in the right direction with the 428 Cobra Jet for 1968. In 1969, Ford unveiled a new Mach 1 SportsRoof that injected a muscle car image into the Mustang, one that was especially fitting for the new performance big-block.

A complete package, the Mach 1 came with hood blackout, scoop, and racing-style click-pins; side and trunk stripe decals (with the first-time use of a reflective material); quad exhaust tips; chrome styled steel wheels; and a pop-open gas cap. But the Mach 1 was more than looks; the package also included the Competition Suspension and a unique Deluxe interior with high-back bucket seats, red accents, woodgrain trim, and a console. Engine availability was limited to 351 cubic inches or larger. For 1970, the Mach 1 was amended with grille driving lights and aluminum rocker panel covers. The combination of the Mach 1 with the optional 428 Cobra Jet was a match made in muscle car heaven—a brute of a 335-horsepower big-block blended with a Mustang named for the speed of sound.

The Mach 1 continued into 1971 on the Mustang's new and larger SportsRoof. Visuals such as the honeycomb grille with driving lights, two-tone paint, and new NACA-duct hood scoops were part of the package, although the special Mach 1 interior moved to the option list. The standard 1971 Mach 1 engine was the anemic 302 two-barrel, so buyers needed to check off one of two 370-horsepower 429 Cobra Jets, including J-code with Ram-Air, to put muscle into the Mach. The big-block disappeared after 1971, so the 1972–73 Mach 1 soldiered on with the 266-horsepower (net rated) 351 Cleveland four-barrel as the most powerful engine.

The 1974–78 Mustang II years were not good to the Mach 1, identified by black lower body side paint but powered by a standard V-6 in 1974 with the 302 two-barrel added as an option for 1975–78. In 1976–78, the Mach 1 was overshadowed by the Shelby-look Cobra II.

mach 1

EVERY MAN'S SPORTS CAR
BOSS

KEY PERSON

Stylist Larry Shinoda arrived at Ford in 1968, just in time to influence the graphics for the Boss 302. His stripes, spoilers, and rear window slats for the 1969–70 Boss 302s were considered radical at the time. He also suggested the "Boss" name.

BY THE NUMBERS

11,796: Total number of Boss Mustangs produced for 1969–71: 8,642 Boss 302s, 1,349 Boss 429s, and 1,805 Boss 351s

TECHNOLOGICALLY SPEAKING

The Boss 429 used cylinder heads with hemispherical-shaped combustion chambers, a more efficient, power-producing design for racing. In addition to their immense size, the Boss 429 hemi heads can be identified by spark plugs mounted in the center of the valve covers.

From 1969–71, Ford produced three Boss Mustangs—Boss 302, Boss 429, and Boss 351, all SportsRoofs with high-performance engines and heavy-duty equipment. Both the 302 and 429 were created to legalize the special high-output engines for racing.

For Trans-Am, the Boss 302 was equipped with canted-valve, large port heads from the four-barrel 351 Cleveland, a solid-lifter cam, and an aluminum intake with a Holley four-barrel carburetor for 290 horsepower. Introduced in April 1969 as a special model, the Boss 302 came with four-speed only, 9-inch rear end supported by staggered shocks, and Competition Suspension with 15-inch wheels and Goodyear Polyglas tires. Ford stylist Larry Shinoda penned the graphics—C stripes and blackout for 1969 and wild wraparound stripes for 1970.

The Boss 429 was Ford's new hemi-head NASCAR engine. Instead of legalizing the racing big-block in the aerodynamic racing Talladega Fairlane, Ford went to extreme lengths to drop the wide Boss 429 into the 1969–70 Mustang, widening the engine compartment with unique inner fenders and sending semicomplete SportsRoofs to Kar Kraft, Ford's contracted performance shop, for the engine installation and final assembly. Rated at 375 horsepower, the most powerful Mustang to date, the Boss 429 was prepped with a mandatory four-speed, 9-inch axle, and Competition Suspension with 15-inch Magnum 500 wheels. Unlike the wildly striped Boss 302, the Boss 429 looked more like a base Mustang SportsRoof; identification was limited to small fender decals and a large hood scoop.

When Ford pulled out of racing in August 1970, there was no longer a need to produce a special Boss 302. Freed from the SCCA's Trans-Am displacement restriction, Ford developed a high-performance, solid-lifter version of the 351 Cleveland. Rated at 330 horsepower, the Boss 351 SportsRoof put more cubic inches under the four-barrel Cleveland heads, resulting in more power and improved torque for an overall improved driving experience.

PERFORMANCE
5.0-LITER HIGH OUTPUT

More than a decade after the last of the big-block Mustangs, and after two years of woefully inadequate 4.2-liter V-8s, performance returned to Mustang with 1982's high-output 5.0-liter. Like 1968's 428 Cobra Jet, Ford engine engineers dipped into the parts bin to revitalize the 302-cubic-inch V-8 with a 351 camshaft, higher flow cylinder heads, a 369-cfm two-barrel carb on an aluminum intake, and a dual snorkel air cleaner for 157 horsepower, the most powerful Mustang engine since 1973. The 5.0-liter was a perfect match for the new Mustang GT, inspiring *Motor Trend*'s cover line, "The Boss is Back!"

Over the next three years, the 5.0-liter matured as a performance engine, gaining a four-barrel Holley for 175 horsepower during 1983–84 and jumping to 210 horsepower in 1985 with a roller camshaft and tubular headers. For 1986, the 5.0's carburetor was replaced by fuel injection for 200 horsepower. The following year, the return of better-breathing open chamber heads increased horsepower to 225, an output that would prevail from 1987 to 1993 (although a new rating system lowered the number to 205 for 1993).

From 1982 to 1993, the 5.0-liter HO was available in both the Mustang GT (hatchback or convertible) and LX (hatchback, convertible, or hardtop). It was the only powerplant for the GT starting in 1983. The LX 5.0 was popular not only for its lower sticker price but also for its tasteful good looks compared to the 1987–93 GT's aero body.

For 1994's new SN-95 chassis and retro-styled body, the 5.0-liter continued as the standard engine for the GT but was no longer available in the base Mustang. A new intake, required to clear the SN-95's lower hood profile, limited output to 215 horsepower.

The 5.0-liter, in production since 1968 as a 302-cubic-inch pushrod small-block, was replaced by Ford's new overhead-cam 4.6-liter in 1996.

The 5.0-liter Mustang's horsepower and improved Fox-body chassis caught the attention of Steve Saleen in 1984. Like Carroll Shelby, Saleen was a racer who saw the opportunity to fund his competition activities by selling his own high-performance cars. He convinced his sister to order a 1984 Mustang GT hatchback and loan it to Saleen Racing as a concept vehicle. Saleen sold only three 1984 Saleens, but they served to establish a performance formula that would insert him into the thick of the Mustang's reemerging performance image.

Saleen was in the right place at the right time, as he noted himself: "The turbocharged SVO had just come out, the GT had generated a lot of interest, and Ford wanted to show how committed they were to performance. They thought another premium-priced Mustang model would add to that image."

With Ford on board to sell Saleen Mustangs through a network of Ford dealers, Saleen outfitted his hatchbacks with lowering springs, Bilstein struts/shocks, subframe connectors, 15-inch wheels, and tweaked alignment specs, along with race-style seats and an upgraded stereo. He chose an existing aftermarket body kit—front air dam, side skirts, and "spats" in front of the rear wheel wells—then added his own tricolor side stripes and a rear spoiler, a "whale-tail" design that would become a Saleen staple for years to come.

Saleen would go on to produce nearly three thousand Saleen Mustangs from 1984–93, including special models such as the 290-horsepower SSC, 326-horsepower SC, and anniversary SA-10 models. The reengineered and restyled 1994 Mustang provided Saleen with a stronger chassis along with a new palette to design a more purposeful look with a new front fascia, side skirts, and rear wing, plus a new S-351 model with a transplanted 351-cubic-inch engine that made 480 horsepower with a supercharger.

Saleen continues to offer limited-production Mustangs on the new S550 chassis.

PERFORMANCE
SVO

In 1980, Ford launched a new Special Vehicle Operations (SVO) group to support private Ford racers, develop performance parts, and create high-performance production vehicles. The 1984–86 SVO Mustang would showcase the program's capabilities.

SVO developed its Mustang, introduced in November 1983 as a 1984 model, to compete against import sports cars such as the Datsun 280-ZX, Toyota Supra, and Isuzu Impulse. Promoted as being "built by driving enthusiasts for driving enthusiasts," the high-tech SVO was powered by a turbocharged and intercooled 2.3-liter four-cylinder engine with 175 horsepower and stopped by the Mustang's first-time use of four-wheel disc brakes. The suspension incorporated adjustable Koni struts/shocks, new Quadra Shock technology at the rear, and 16-inch wheels mounted by five lugs (instead of four like other Fox-body Mustangs) and carrying Goodyear Eagle performance tires. Externally, the SVO stood out with its sleeker front fascia, foglights, hood with offset and functional intercooler scoop, spats at the rear wheel openings, and a distinctive dual-wing rear spoiler. Inside, the SVO's monochromatic charcoal interior justified the car's $15,000 sticker price with inflatable lumbar support for the sport seats, a leather-wrapped steering wheel, and an AM/FM stereo with Premium Sound. Confirming its status as a driver's car, the SVO also came with a Hurst shifter, a wide brake pedal to assist with heel-and-toe shifting, and a dead pedal foot rest to aid driver bracing during hard cornering.

For 1985, SVO acceleration benefitted from 3.73 gearing. At midyear 1985, Ford upgraded the SVO's engine with a new camshaft, reworked intake, split exhaust system, and one-pound boost increase for 205 horsepower, improvements that continued into 1986.

Regardless of the SVO's modern technology and all-around performance, most Mustang buyers preferred the torque—and lower price—of the 5.0-liter V-8. Ford sold fewer than ten thousand SVO Mustangs during its three-year run.

PERFORMANCE
COBRA

Similar to SVO in the 1980s, Ford's Special Vehicle Team (SVT) was tasked with marketing performance parts and vehicles into the 1990s. SVT quickly outpaced SVO by producing several high-performance Fords, including Mustang Cobras and Lightning F-150 pickups.

SVT's first Mustang was the 1993 Cobra, introduced during the final months of the Fox-body era. The Cobra lived up to its legendary name with 235 horsepower, 30 more than the standard 5.0-liter HO thanks to GT-40 heads, a unique intake, and a more aggressive camshaft. The extra power was supported by an upgraded suspension with 17-inch Goodyear Eagle tires. Visually, the Cobra differed from GTs with a narrow grille opening, a unique rear spoiler, and vane-spoked aluminum wheels.

Shortly after the introduction of the new 1994 Mustang, SVT launched its latest version of the Cobra, once again with more power (240 horsepower, the most ever for a production 5.0-liter to date), a bumper cover with round driving lights, larger disc brake rotors, and a restyled rear spoiler. It was also offered as a convertible, which was chosen as the pace car for the 1994 Indianapolis 500.

The Cobra took a huge leap forward in 1996. The new dual-overhead-cam (DOHC), four-valve 4.6-liter modular V-8 developed 305 horsepower, once again a Mustang high-water mark, as SVT's Cobra became more of a mainstream offering alongside the GT. In 1999, the Cobra was the first Mustang equipped with an independent rear suspension. The Cobra skipped 2000 entirely while SVT sorted out an engine problem, and the 2002 model was canceled when power didn't meet Chief Engineer John Coletti's expectations. However, the resulting 2003 Cobra was worth the wait. An Eaton supercharger boosted horsepower to 390, yet another record for Mustang and one that was bolstered by a new six-speed manual transmission, upgraded suspension, and a new front fascia for improved engine cooling.

The Cobra name was retired after 2004 and replaced in 2007 by the Shelby GT500.

PERFORMANCE
MODULAR DOHC

TECHNOLOGICALLY SPEAKING

To exploit the capabilities of the four-valve engine without sacrificing low-end torque, the DOHC 4.6 used a port throttle system. Below 3,250 rpm, air and fuel were delivered to the primary intake valves. At higher rpm, electronically controlled secondary valves opened.

TOP QUOTE

"Enthusiasts have figured out that if you really want to increase power output from a Ford modular engine, you install a blower." —John Coletti, describing the 2003 Cobra's supercharged engine

FUN FACT

The 2003–04 Mach 1's hood scoop was a duplicate of the Shaker scoop found on 1969–70 Ram-Air Mustangs. Ducting fed cooler outside air from the scoop to the factory air cleaner assembly.

The Mustang rolled into the modern powertrain era for 1996 with a new overhead-cam 4.6-liter engine, described as "modular" because it could be built in a variety of cylinder configurations on the same assembly line. While the Mustang GT was powered by a two-valve 4.6, the SVT Cobra came with a dual-overhead-cam (DOHC) version sporting four valves per cylinder for 305 horsepower. Sharing only a few components with the two-valve, the DOHC's bottom end used a lightweight aluminum block with a forged crankshaft and powder-metal connecting rods. On top were four-valve heads with a pair of cams for each bank. A multirunner intake with secondary throttle plates, combined with more aggressive timing, allowed the engine to make horsepower and torque up to its 6,800-rpm rev limit. Final assembly of each Cobra DOHC 4.6 was handled at Ford's Romeo engine assembly plant by twelve teams with two technicians per engine. A plate on a valve cover included the signatures of the engine builders.

Realizing that the DOHC 4.6-liter had reached its maximum horsepower potential as a naturally aspirated engine, SVT chief engineer John Coletti urged his powertrain team to explore supercharging, a popular aftermarket power-adder for Mustang hot-rodders. The Eaton Roots—style blower offered a quick path to Coletti's power goal. Bolstered by revised aluminum heads and a water-to-air intercooler, the 2003–04 Cobra's supercharged DOHC 4.6 generated 390 horsepower and 390 pound-feet of torque. To handle the added stress, the engine was built from a cast-iron block with Manley H-beam connecting rods, forged pistons, and an aluminum flywheel.

The naturally aspirated DOHC 4.6 made a special reappearance in the 2003–04 Mach 1. Topped by a retro Shaker hood scoop, the Mach's 305 horsepower slipped perfectly between the GT and Cobra.

PERFORMANCE
ROUSH

KEY PERSON

Jack Roush bought a new 1965 Mustang shortly after landing his first job with Ford in 1964 and succeeded in drag racing before launching his own company. Today, Roush Industries employs over 2,700 people at forty facilities, including Roush Performance.

HISTORICAL TIDBIT

For 1989, Ford asked Roush to develop a Mustang 25th anniversary model, resulting in a hatchback powered by a twin-turbocharged 351. Ford described the power as "irresponsible," and the car was not approved. Today, the twin-turbo Mustang is part of Roush's museum.

FUN FACT

Jack Roush is an avid aviation enthusiast. His P-51 Roush Mustangs are named for the famous World War II P-51 Mustang fighters and carry aviation-like graphics.

After a successful drag racing career with Wayne Gapp and winning numerous Pro Stock championships in Gapp and Roush Fords, Jack Roush left his job at Ford to launch Jack Roush Performance Engineering (now Roush Performance). Combining his racing experience with engineering savvy, Roush's company provided services not only to racers but also to the Big Three auto manufacturers. In 1984, Ford encouraged Roush to return to racing, only this time for the SCCA and IMSA road courses. Roush-prepared Fords would win numerous championships, including ten consecutive IMSA GTO wins, most with Mustangs, at the 24 Hours of Daytona.

Roush Performance built its first street production Mustang in 1995 as a 5.0-liter with a Roush-designed intake and cowl-induction hood. With Ford's 1996 switch to the 4.6-liter, Roush began offering Stage 1, Stage 2, and Stage 3 Mustangs, starting with exterior enhancements and progressing up to the Stage 3's all-out performance package with suspension upgrades and a supercharger. By 2002, Roush had added Sport, Rally, and Premium options to the Stage 3.

During the 2005–14 S197 era, Roush continued with the Stage models but also produced many special editions, including the 427R, P-51A and P-51B, BlackJack, 428R, 440A, 380R, and RTC. A Roots-type ROUSHcharger supercharger with a water-to-air intercooler, a new aluminum intake, and exclusive Roush calibration generated between 417 and 675 horsepower for the Stage 3 and other supercharged models. Thanks to Roush's status as an OEM supplier, the Roush Mustangs were known for their quality.

Roush continues building Stage 1, 2, and 3 Mustangs on the latest 2015–18 Mustangs. All include Roush body components and graphics. The Stage 1 is now based on the EcoBoost four-cylinder, while the Stage 3 pushes Coyote 5.0-liter horsepower to 670 with a Roush 2300 TVS supercharger.

PERFORMANCE
5.0 COYOTE

For 2010, Ford gave the Mustang a leaner and meaner facelift; for 2011, they gave the GT a major horsepower boost with a modern 5.0-liter four-valve V-8 with Twin Independent Variable Camshaft Timing (Ti-VCT) for 412 horsepower, a 97-horsepower improvement over the outgoing 4.6-liter from 2010. Codenamed "Coyote," the new engine revived the famous 5.0-liter engine displacement from the Fox-body era but otherwise had nothing in common with the older pushrod V-8.

The Coyote's aluminum block was developed for optimized high-rpm use during enthusiast track-day outings. The aluminum four-valve-per-cylinder heads were also new, adding a compact roller-finger follower valvetrain layout to provide more room for higher-flow ports. Output would be adjusted to 420 horsepower for 2013–14.

The new 5.0-liter's 302-cubic-inch displacement allowed Ford to bring back the Boss 302 for 2012–13 as a track-ready Mustang with 444 naturally aspirated horsepower.

Using lessons learned from the modern Boss 302 development, the 5.0-liter in the all-new 2015 Mustang GT generated 435 horsepower. *Car & Driver* magazine reported a quarter mile elapsed time of 13.0 seconds at 113 miles per hour with the six-speed manual transmission, easily topping the best performing Mustang muscle cars from the 1960s.

Ford continued to refine the Coyote 5.0-liter into the Mustang's 2018 model year, describing it as "totally reworked" with revised camshafts and new cylinder heads for 460 horsepower and 420 pound-feet of torque, vaulting the Mustang ahead of the Camaro SS in the modern muscle car war. A combination of direct- and port-fuel injection, along with a higher compression ratio, contributed to the increase, which propels the 2018 Mustang GT from 0 to 60 in under 4 seconds when equipped with the new ten-speed automatic transmission and available Drag Strip mode.

SURVIVAL OF THE FASTEST

Performance has always been important to Mustang. Early on, it was about image. High-performance models never sold in high numbers during the first generation, but intrigued customers would stroll into a Ford dealership to inspect the new Shelby or Boss and drive out in a six-cylinder hardtop with a three-speed stick on the floor. During 1974 to 1981, there was little substance under the hood, but the performance image still carried on with the Mach 1, Cobra II, and King Cobra models. Thanks to modern technology, today's Mustangs offer a combination of performance, smooth operation, and fuel economy.

Competition with Chevrolet's Camaro often drove Mustang performance to new heights. It was the new 1967 Camaro's available 396-cubic-inch big-block that forced Ford into offering the 390 in the 1967 Mustang. The 1967–69 Camaro Z/28 also inspired the 1969–70 Mustang Boss 302, a Chevy versus Ford battle that spilled over from the street and onto the Trans-Am race tracks. The Z/28 was king in the late 1970s and early 1980s but was later dominated by the 1987–93 5.0-liter Mustang Fox-body. Mustang claimed the ultimate victory in 2002 when the Camaro was discontinued, though it would return for 2010. The war rages on today as both Ford and Chevy seesaw horsepower ratings with SS Camaros and GT Mustangs.

Over the years, Mustang performance has progressed from special models such as Shelby and Boss to more mainstream models with the GT and EcoBoost, putting high horsepower within reach of more enthusiasts. Ford continues to add spice to the mix with special models such as the 2012–13 Boss 302,

2007–14 Shelby GT500, and 2015–17 Shelby GT350. And for those looking to take performance to the next level, companies such as Shelby
American, Roush Performance, and Saleen offer Super Snake, Stage 1,
and Black Label Mustangs.

GLOSSARY: PERFORMANCE

AMA: The Automobile Manufacturers Association, a trade group that proclaimed a ban on factory-supported racing and performance in 1957 following incidents where race cars crashed into grandstands.

Boss: Youth slang for something cool or great in the late 1960s; used for special performance Mustangs in 1969 to 1971 and 2012–13.

Cleveland: The designation for the Mustang's mid-displacement V-8 engine from 1970 to 1973, as used to differentiate the newer canted-valve powerplant from the earlier inline-valve Windsor version.

Cobra Jet: The name for the higher-performance 428s and 429s that were optional for Mustangs from 1969 to 1971.

Coyote: Ford's internal code for the overhead-cam, four-valve 5.0-liter V-8 engine that debuted in the Mustang GT for 2011 with 412 horsepower. By 2018, the horsepower rating had improved to 460.

Displacement: Measured in cubic-inches or liters, the total volume of air/fuel mixture that an engine can draw in during one complete engine cycle. It is calculated by multiplying the number of cylinders by the bore and stroke.

Hi-Po: Short for 289 High Performance, this is Ford's high-output small-block engine, equipped with four-barrel carb, solid lifters, and high-flow exhaust manifolds for 271 horsepower.

Mach 1: By definition, equal to the speed of sound. The name of a performance image Mustang model offered from 1969 to 1978 and 2003–04.

Ram-Air: In the late 1960s and early 1970s, this was Ford terminology for cold-air induction as a method for drawing cooler outside air into the air cleaner via hood scoops or ducts.

Supercharger: A belt-driven induction system that forces more air and fuel into an engine for increased horsepower. From 2007 to 2014, the Shelby GT500's roots-style supercharger helped produce from 500 to 662 horsepower.

SVO: Special Vehicle Operations, a Ford division that developed and sold performance parts in the 1980s. SVO also developed and marketed the high-performance SVO Mustang from 1984 to 1986.

SVT: The Special Vehicle Team, created by Ford in 1993 as a successor to SVO. It developed and marketed high-performance 1993 to 2004 Cobra Mustangs, 2007 to 2014 Shelby GT500s, and 2015 to 2018 Shelby GT350s.

Voodoo: A 5.2-liter variation of the Coyote V-8 engine with flat-plane crankshaft technology, as used in the 2015 to 2018 Shelby GT350.

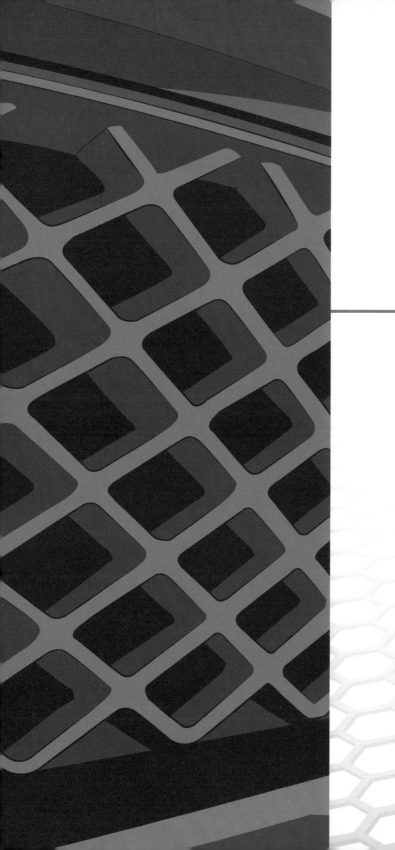

SHELBY

SHELBY
CARROLL SHELBY

Army Air Force pilot. Chicken farmer. Race car driver. Car builder. Le Mans champion. Big-game hunter. Car builder again. Over his eighty-nine years, Carroll Shelby lived a full and exciting life. He rubbed shoulders with celebrities, courted beautiful women, and traveled the world. For Carroll Shelby, the Mustang was a small slice of a life well lived. But for Mustang, Shelby was a major contributor to its performance legacy.

Shelby's name is widely recognized for Ford race cars, Cobra sports cars, and fast Mustangs. But before that, in the 1950s, he was a successful race car driver, competing internationally for sports car manufacturers such as Aston Martin, Ferrari, Porsche, and Maserati. *Sports Illustrated* named Shelby its Driver of the Year in 1956 and 1957. He then capped his driving career by winning the 1959 24 Hours of Le Mans. However, after the 1960 season, the 37-year-old Shelby retired from driving due to a heart ailment.

Shelby then turned his attention from driving cars to building them, running full throttle with an idea to power a European sports car with a small American V-8. Through his contacts, Shelby learned two timely and important facts: England's AC Ace was losing its engine supplier and Ford was introducing a compact and lightweight small-block V-8. By the end of 1962, Shelby was building, selling, and racing the Ford-powered Cobra, which added the sports car piece to Ford's Total Performance puzzle. Shelby American quickly became Ford's go-to company for road racing.

In 1964, when Ford's Lee Iacocca wanted to project more of a performance image for the Mustang, he called Shelby. Although busy with Cobra and racing projects, Shelby accepted the challenge, one that would forever link the Shelby name to the Mustang's legacy. Shelby GT350s were built from 1965—with GT500s added in 1967—until Shelby himself pulled the plug in 1969. He would return to Ford in 2005 to start a new era of Shelby Mustang performance.

SHELBY
GT350

Conferring with the Sports Car Club of America (SCCA), Carroll Shelby discovered that he could transform the Mustang into the SCCA's definition of a sports car by removing the rear seat, effectively making it a two-seater. Starting with Wimbledon White 1965 fastbacks equipped with the 289 High Performance engine, Shelby American replaced the rear seat with a fiberglass panel, then enhanced the performance with suspension upgrades and engine modifications, taking horsepower from 271 to 306. Loud side exhaust was part of the fun, and the optional over-the-top Guardsman Blue "Le Mans" stripes would become a Shelby staple for years to come. Only 562 GT350s were produced for 1965, including thirty-three Competition versions, also known as "R-Models," that would help Shelby American win the SCCA B-Production championship.

Looking to increase sales, Shelby American tamed the 1966 GT350 by eliminating the side exhaust and locking rear end, retaining the back seat, and offering more exterior colors and an automatic transmission. Side brake cooling scoops and plexiglass windows in place of the fastback's C-pillar vents helped differentiate the Shelby from the standard Mustang. Bolstered by Hertz Rent a Car's order for one thousand GT350H models, second-year Shelby sales jumped to 2,378.

Restyled on the larger 1967 Mustang, the revamped GT350 looked more like a Shelby and less like a Mustang with more extensive use of body fiberglass, grille-mounted high-beam headlights, and wide Cougar taillights. A big-block GT500 joined the GT350, which continued with the Cobra version of the 289 High Performance. For 1968, the Shelby transitioned from all-out performance to more "Road Car" luxury with a plush interior and power from a smooth 302-cubic-inch small-block rated at 250 horsepower.

For 1969, the Shelby once again adapted to the Mustang's new styling, this time with even more fiberglass for a totally unique and fresh look. The GT350 was powered by the 290-horsepower 351-cubic-inch Windsor small-block.

SHELBY
GT350 HERTZ

The 1966 GT350 got a boost, both in sales and in promotion, when Hertz ordered Shelby Mustangs for its Hertz Sports Car Club, which had rented out Corvettes and other two-seaters before switching from GM to Ford in 1965. Shelby American sales manager Peyton Cramer approached Hertz with the idea of a GT350H, thinking it might result in a few extra sales. Building the prototype in black and gold, the Hertz corporate colors, didn't hurt the sales pitch. Hertz initially implied it would order five hundred to seven hundred cars. Eventually, it would take delivery of one thousand.

Prequalified Hertz Sports Car Club members could rent a Shelby GT350H for $13 a day and 13 cents a mile. Renters spent the weekend with a Mustang thrill ride that included the usual GT350 performance amenities—a 306-horsepower 289 Cobra engine, adjustable Gabriel shocks, a fiberglass hood with scoop, a dash-mounted tachometer, competition seat belts, 14-inch Magnum 500 wheels, metallic brake pads and shoes, and rear quarter windows. Most were equipped with an automatic transmission. To warn renters about the brake feel, especially when cold, a small decal was placed on the instrument panel: "This vehicle is equipped with competition brakes. Heavier than normal brake pedal pressure may be required." Shelby American worked with Hertz to solve the brake feel complaints, adding a power booster to some cars and later replacing the hard racing pads with softer versions.

The Hertz Shelby GT350s are generally recognized by their black-and-gold paint scheme. However, approximately two hundred were built in red, green, white, and blue, all with gold stripes.

After approximately a year in the Hertz rental fleet, the Shelby GT350H Mustangs were returned to Shelby dealers, reconditioned, and sold as used cars. Due to their interesting history, the Hertz Shelbys are among today's most desirable collector cars.

SHELBY
GT500

BY THE NUMBERS
500: Like the "350" in GT350, the "500" in GT500 didn't represent engine displacement or horsepower. "It just sounds bigger than anything else" was the Shelby American explanation.

TOP QUOTE
"It's a beautiful thing. Those bulging lines you see from the driver's seat give you a sense of power and a stomp on the throttle pedal will tell you it's not just wishful thinking." — *Car Life* magazine, describing the 1969 Shelby GT500

FUN FACT
For the 2001 remake of the 1973 film *Gone in 60 Seconds*, the "Eleanor" star-car was dressed as a 1967 Shelby GT500. The muscled-up fastback brought new attention to the Shelby Mustang.

The availability of the 390 engine in the 1967 Mustang presented Shelby American with the opportunity to offer a big-block version of its Shelby Mustang. While the GT350 continued with the "Cobra" 289 High Performance, the new GT500 took a step in the muscle car direction by replacing the Mustang's 390 with a 428 Police Interceptor—but not just any 428 PI. The Shelby version was topped by an aluminum medium-riser intake with a pair of 600-cfm Holley four-barrels, making an impressive sight under the fiberglass hood along with the oval finned aluminum air cleaner and tall "Cobra Le Mans" valve covers. Horsepower was rated at 355. Both four-speed and automatic were available, as was air conditioning for the first time in a Shelby.

For 1968, the mildly restyled GT500 lost its dual-quad induction, replaced by a single four-barrel version that was strangely rated at 5 more horsepower than the previous year's twin Holley engine. Like the GT350, the 1968 GT500 was available as a convertible for the first time and included a unique roll bar. In April 1968, the 428 PI was replaced by Ford's new 428 Cobra Jet, resulting in the GT500KR name.

The 428 Cobra Jet continued as the standard powerplant for the 1969 Shelby GT500, which was available in both SportsRoof and convertible. As an R-code CJ, the single Holley carburetor drew cooler outside air from the center hood duct, while the other four openings contributed to lower under-hood temperatures. When ordered with the optional Drag Pack, the GT500 came with 3.91 or 4.30 gears, and the 428 was upgraded to Super Cobra Jet status with an oil cooler and strengthened bottom end.

SHELBY
GT500 SUPER SNAKE

When Goodyear requested a 1967 Shelby GT500 for a high-speed tire test, Carroll Shelby asked his chief engineer, Fred Goodell, to prepare a special test vehicle. Choosing a white production GT500, Goodell instructed Shelby American technicians to prepare and install a medium-riser 427 with aluminum heads, a solid-lifter cam, a 780-cfm Holley four-barrel, GT40-style headers, and an oil cooler. A four-speed and Detroit Locker 4.11 rear axle completed the package. To differentiate the specially prepared Shelby, the Guardsman Blue Le Mans stripes had a triple narrow-wide-narrow pattern instead of the usual dual stripes.

The unique 427-powered Shelby Mustang would become known as the GT500 Super Snake.

For the tire test, the Super Snake was delivered to Goodyear's 5-mile oval test track in San Angelo, Texas, and shod with Goodyear's new Thunderbolt tires. With both *Time* and *Life* magazines covering the event, Carroll Shelby showed up to take journalists on rides topping 150 miles per hour. Later, Fred Goodell established a new record by averaging 142 miles per hour for 500 miles.

Record set and purpose over, the GT500 Super Snake was shipped back to Shelby American and offered for sale to Shelby dealers. However, the 427-powered Shelby Mustang sparked an idea in Don McCain, a former Shelby American western sales representative who had gone to work as performance sales manager at Mel Burns Ford. McCain wanted to market a special 427 Super Snake GT500, similar to the 427-powered Camaros offered by the Yenko and Nickey Chevrolet dealerships. McCain proposed building fifty Super Snakes, but the idea hit a roadblock when he realized that the costs of the 427 would send the sticker price soaring to over $7,500.

Although Shelby American built two other 427-powered 1967 GT500s, the Goodyear test car would go down in Shelby history as the only 1967 GT500 Super Snake.

SHELBY
GT500KR

In April 1968, the Mustang's new 428 Cobra Jet big-block replaced the GT500's 428 Police Interceptor. To differentiate the updated CJ-powered cars from the earlier models, "KR" was added to the GT500's side stripes for "King of the Road."

Carroll Shelby came up with the name. He told the story many times, including this version: "Lee Iacocca had an assistant named Hank Carlini. And Hank was kind of the spy who kept on what Chrysler and General Motors were doing. And he said, 'Did you know that Corvette is coming out with a King of the Road? They're fixed to announce it in two weeks and here's the brochure.' So I called my trade-dress lawyer in Washington—it was about three o'clock in the afternoon—and said, 'I want to know if King of the Road has been copyrighted.' He said he'd find out in the morning. I said, 'I'll have another trade-dress lawyer by the morning. I want to know now.' He called me back in about an hour and said it wasn't taken. I said, 'You better be there in the morning to take it.' He did. I called 3M to make the decals with GT500KR."

In addition to the 428 Cobra Jet engine, the GT500KRs were updated with braced shock towers. They were available as automatic or four-speed, with manual transmission cars also equipped with staggered rear shocks to dampen wheel hop during hard acceleration.

Today, the Shelby GT500KR fastbacks and convertibles are highly prized by collectors. Restored and survivor models command well over six-figure sale prices, placing them among the most valuable of all Shelby Mustangs.

In 2008–09, Shelby American revived the GT500KR name for a 540-horsepower Shelby Mustang that was sold through Ford dealerships.

SHELBY
MODERN GT500

HISTORICAL TIDBIT

Carroll Shelby worked closely with SVT during the 2013–14 GT500 development, making it the last Shelby Mustang that he contributed to before his death in May 2012.

BY THE NUMBERS

202: Top speed for the 2013 Shelby GT500 as reported by SVT test drivers at the Nardò Ring, Italy's 12.5-mile test track

TECHNOLOGICALLY SPEAKING

The 2013 Shelby GT500 was the first production Mustang to come from the factory with a carbon-fiber driveshaft.

Three decades after Carroll Shelby dismantled his Shelby Mustang program in 1969, the rumor was intriguing: Carroll was once again roaming the halls at Ford Motor Company. In 2005, the gossip was confirmed when Shelby appeared onstage at the New York International Auto Show to introduce the 2007 Shelby GT500.

Based on the S197 Mustang with its retro styling, the modern GT500 was actually the latest iteration of the SVT Cobra, rebranded as a Shelby to rekindle the legacy of the top performance Mustang from the 1960s. Only it was better than the 1960s. SVT touted the 2007 Shelby GT500 as "the most powerful factory-produced Mustang ever" thanks to a Roots-style supercharger atop a thirty-two-valve 5.4-liter engine. The 500 horsepower added true meaning to the GT500 name. SVT engineers also enhanced the new Shelby with a six-speed transmission, a race-tuned suspension, and Brembo brakes. Externally, the GT500 combined styling cues from both SVT and Shelby, including an aggressive front end that mimicked the 1968 Shelby.

For 2010, SVT updated the GT500's styling and increased horsepower to 540. But they saved the best for last. While the 2013–14 Shelby didn't look much different from the previous 2010–12 models, it was what you couldn't see that made the latest GT500 the most impressive production Mustang ever. Under the hood, the 5.4 was replaced by a 5.8-liter topped by a higher capacity supercharger for 14 pounds of boost. At the time, the 2013–14 GT500's 662 horsepower made it the most powerful production V-8 in the world. With 3.31 gearing, the GT500 also avoided the gas guzzler tax with fuel ratings of 15 miles per gallon city and 25 miles per gallon highway.

SHELBY
MODERN GT350

Speculation ran rampant when Ford introduced the all-new 2015 Mustang without an SVT variant. Would the Shelby return later as an even more powerful GT500? Or would SVT debut something totally different?

The answer came in December 2014 at Carroll Shelby's warehouse and office facility in Gardena, California. "Here we are in Carroll's house," said Ford Executive Vice President Jim Farley as he introduced the white-with-blue-stripes 2016 Shelby GT350.

Taking a name from Shelby's past and breaking from the previous GT500's supercharged reputation as a high-horsepower road car, the modern GT350 was developed as the most technologically advanced Mustang ever offered and, per the press release, was "capable of tackling the world's most challenging roads and race tracks." To that end, the 2016 Shelby GT350 was powered by an all-new 5.2-liter V-8 with a flat-plane crankshaft, as used in racing exotics such as the Ferrari 458, Porsche 918 Spyder, and McLaren P1. Combined with cylinder head and valvetrain improvements, the new engine redlined at 8,200 rpm. It was rated at 526 horsepower to go down in history as the most powerful naturally aspirated Ford ever offered to the public.

With handling already advanced thanks to the S550 Mustang's independent rear suspension, the GT350 took another step forward with the first-ever use of MagneRide dampers, which utilized wheel-positioning sensors to change stiffness and dampening characteristics within milliseconds. The package also included Brembo brakes and 19-inch wheels with Michelin Pilot Super Sport tires. A wider front track mandated reshaped front fenders, which incorporated vents to draw out turbulent wheelwell air. To increase downforce at speed, the GT350's front bodywork was 2 inches lower than the regular Mustang GT.

As expected, the 2016 Shelby GT350 was popular with Mustang and track-day enthusiasts. It continued into the 2017 model year.

THE SHELBY LEGACY

Carroll Shelby passed away on May 10, 2012, at age 89, leaving a legacy that spread far beyond Mustangs.

Shelby was named *Sports Illustrated*'s Driver of the Year in 1956 and 1957. As a driver, Shelby won the 24 Hours of Le Mans in 1959. Later, he would also win at Le Mans as a manufacturer (1964, Cobra Daytona Coupe) and team owner (1966, Ford GT 40), becoming the only person to accomplish that trifecta.

After ending the Shelby Mustang program in 1969, Carroll spent much of the 1970s hunting in South Africa.

During the 1980s, Shelby returned to the automotive industry to help his old friend Lee Iacocca at Chrysler. Shelby added his magic to create the Shelby Dakota pickup and the front-wheel-drive Shelby Charger, Daytona, and Omni, which was offered in two models—the GLH (for "Goes Like Hell") and the supercharged GLH-S (for "Goes Like Hell Somemore").

As a Texan, Shelby liked chili. In 1967, he created a chili cook-off in Terlingua, Texas, which spawned the International Chili Society. His Carroll Shelby's Chili mix is still available today. Shelby's chili cook-offs also inspired his son-in-law at the time, Larry Lavine, to start the Chili's restaurant chain.

In the 1990s, Shelby established Shelby Automobiles to build continuation Cobras and the two-seater Shelby Series 1, a car that wasn't based on an existing chassis, making it the only vehicle built by Shelby from the ground up.

Shelby survived two organ transplants, heart and kidney. While recovering from his heart transplant, he became aware of the hardships experienced by children needing a transplant and created the Carroll Shelby's Children's Foundation in 1991. Today, it is known as the Carroll Shelby Foundation, a charity organization that assists children from childhood through their school years.

GLOSSARY: SHELBY

Eleanor: A Shelby-look 1967 Mustang fastback created for the 2001 movie *Gone in 60 Seconds*.

GT350: The original Shelby Mustang, powered by small-block 289, 302, or 351 engines from 1965 to 1970. It returned in 2015 as SVT's track version of the Mustang.

Gas guzzler: A vehicle that is perceived to use a lot of fuel and a description that came into common use when Congress established a Gas Guzzler Tax in 1978 to discourage fuel-inefficient cars.

GT500: A Shelby Mustang produced from 1967 to 1970 with 428 big-block power, then returning from 2007 to 2014 as a high-performance model from Ford SVT.

KR: King of the Road, a name originally slated for a performance Chevrolet but copyrighted first by Carroll Shelby for the Cobra Jet–powered 1968 GT500KR. It returned in 2008–09 as a more powerful version of SVT's GT500.

MagneRide: An adaptive automotive suspension, as used on the 2015 to 2018 Shelby GT350, that utilizes magnetically controlled dampers to immediately change suspension settings based on input from steering, acceleration, and other factors.

Plexiglass: A transparent plastic material used as a lighter-weight alternative for glass. Used for the unique rear quarter windows in the 1966 Shelby GT350s.

Rent-a-racer: The Shelby Mustangs produced for and rented out by Hertz Corporation in 1966, 2006, and 2016.

R-Model: An unofficial name coined by enthusiasts for the 1965 GT350 competition model.

Road car: A marketing description for the 1968 Shelby Mustang to support its new appeal as a high-end sports car with luxury, as opposed to the "boy racer" image of the earlier Shelbys.

SCCA: The Sports Car Club of America, a race sanctioning body formed in 1944 that supports road racing and autocross. Over the years, Mustang succeeded in its B-Production and Trans-Am classes.

Super Snake: A special 1967 GT500 powered by a 427 and built as a high-speed vehicle for a Goodyear tire test. A proposal to make it a unique model was turned down when the retail cost was deemed too high.

SPECIAL EDITIONS

SPECIAL EDITIONS
PACE CAR REPLICA

The Mustang had been available for only a few months when Ford created the first unique model. Unlike future special editions that were designed primarily as sales enticements, the 1964½ Indy Pace Car replica showcased the Mustang's selection as the pace car for the 1964 Indianapolis 500 while also serving as a dealer sales reward.

Pacing the Indianapolis 500 was a big deal for Ford in the early days of Mustang mania. For the actual race, Holman Moody prepared three convertibles with 289 High Performance engines (not yet available to the public in the Mustang), while Ford provided thirty-five convertibles in Fleet White to transport the beauty queens during the prerace parade laps. All had blue Rally stripes and "Official Indianapolis 500 Pace Car—Ford Mustang" lettering.

Taking advantage of the publicity, Ford produced Pace Car replica hardtops for a pair of dealer sales contests, with the top five performing dealers in each sales district receiving a car. Approximately 190 were built, with keys to many presented by Lee Iacocca during a "Checkered Flag" presentation in Dearborn where the dealers took delivery of their prizes.

All Pace Car replica hardtops were painted in a special code C Pace Car White. Each was equipped with white interior with blue appointments, a 260-cubic-inch V-8, and an automatic transmission. The "Official Indianapolis 500 Pace Car" decals and over-the-top Rally stripes were also part of the package. Some of the cars were displayed in dealership showrooms or driven as dealer demonstrators. They were eventually sold and used as daily drivers. Most were eventually lost to the ravages of mileage and time. Today, the 1965 Mustang Pace Car replicas are considered unique and interesting collector cars.

Mustang was also selected as the Indy 500 pace car in 1979 and 1994, resulting in special replica editions.

SPECIAL EDITIONS
SPRINT

By the spring of 1966, Mustang sales were reaching the saturation point with over 1 million sold. Ford celebrated the accomplishment—and spurred late model-year sales—by creating a special model at a low price. Offered in all three Mustang body styles, the Sprint 200 came with the 200-cubic-inch six-cylinder and packaged a number of popular options: an automatic transmission, a console, wire wheel covers, rocker panel moldings, and pinstripes. A unique underhood identifier was a chrome air cleaner with a "Sprint 200" decal. Sprint models were produced from March 1966 until the end of the model year.

The spring promotion returned in 1967 as the Sports Sprint for hardtops and convertibles to add the popular louvered hood with integral turn signal indicators, rocker and tail panel moldings, wheel covers, white-wall tires, a vinyl-covered shifter handle, and a chrome air cleaner. For 1968, the Springtime Sprint was offered in two different packages. "Sprint A" was available with either a six- or eight-cylinder engine and came with side C stripes, wheel lip moldings, wheel covers, and a pop-open gas cap. "Sprint B" was limited to the V-8 and added styled steel wheels with E70 tires and grille-mounted foglights.

After skipping three years, the Sprint returned in the spring of 1972 as a red, white, and blue special edition for Mustang, Maverick, and Pinto. Offered from March to June, the Sprint Décor Option Group supplied a white hardtop or SportsRoof with blue and red trim, including "USA" shield decals on the rear quarters. Package A provided the Sprint exterior, white interior with blue seat inserts and red piping, dual racing mirrors, hub caps with trim rings, and E70 white sidewall tires. Package B added the competition suspension, including F60 tires on Magnum 500 rims.

SPECIAL EDITIONS
CALIFORNIA SPECIAL

In February 1968, California Ford dealers started marketing their own special-edition Mustang, called the "California Special." The idea came from a Shelby styling prototype known as "Little Red," a 1968 Mustang hardtop with the Shelby rear end treatment and side scoops. Looking for a unique model for the all-important California sales district, which sold 20 percent of all Mustangs, Southern California District Sales Manager Lee Grey approached Ford about producing a special hardtop. A GT/SC, for "GT Sports Coupe," that was under development at Ford eventually became the GT/CS for "California Special."

Shelby Automotive's Fred Goodell, who had created the "Little Red" hardtop earlier, got the assignment to oversee the engineering and production of the GT/CS. Two prototypes were hand-built at Shelby American's race shop and presented to California Ford executives for review. While the car was originally intended for southern California, the decision was eventually made to market the GT/CS throughout the entire state. One magazine ad included a photo of Carroll Shelby and a California Special with the headline, "Only Mustang and Carroll Shelby could make this happen!"

The GT/CS was primarily an appearance package, adding Marchal or Lucas foglights in the front grille, twist-type hood locks, side scoops, a pop-open gas cap, and the Shelby rear end with a spoiler and 1965 Thunderbird taillights. All 1968 Mustang exterior colors were available, with contrasting white, black, red, or blue side stripes that ended in die-cut "GT/CS" lettering on the scoops. Most California Specials were powered by the two-barrel 289 and featured an automatic drivetrain, but all Mustang engines were available. A few were built with the 428 Cobra Jet.

Final assembly of the 1968 California Specials took place at Ford's San Jose assembly plant between January 18 and July 18, 1968.

SPECIAL EDITIONS
POLICE CARS

BY THE NUMBERS
15,000: Approximate number of SSP Mustangs produced from 1982–93, all hardtops except for five hatchbacks ordered by the California Highway Patrol in 1982

FUN FACT
Ford promoted the SSP Mustangs in advertising that depicted a photo of a California Highway Patrol hardtop at speed and the headline, "This Ford chases Porsches for a living."

HISTORICAL FACT
The United States Air Force used SSP Mustangs as chase vehicles to assist pilots when landing U-2 reconnaissance planes.

In the early 1980s, police departments turned to the 5.0-liter High Output Mustang as an alternative pursuit vehicle to the full-size Ford Crown Victoria, which had become underpowered due to tightening emissions standards. Disappointed after testing 1979 Camaros, the California Highway Patrol ordered over four hundred 1982 Mustangs, mostly hardtops, with the 5.0-liter engine, four-speed manual transmission, and special equipment that included a calibrated speedometer, single-key locks, and a full-size spare tire.

Recognizing the appeal (and sales potential) of the 5.0 Mustang as a pursuit vehicle, Ford created a Special Service Package (SSP) for 1983 Mustangs, adding a two-piece VASCAR speedometer cable and radio noise suppression to the 1982 CHP package. The SSP would continue through 1993, eventually being utilized by over thirty highway and state patrols, with California, Florida, and Texas employing the highest number. SSP Mustangs were also ordered by city police departments, the FBI, the Drug Enforcement Agency, US Border Patrol, and other federal agencies.

During the SSP's eleven-year production span, Ford adapted the package to fit pursuit needs. From 1986–88, the SSP Mustangs were also equipped with more durable silicone hoses, aircraft-type hose clamps, a reinforced front floor pan, front disc brake rotor shields, a Kevlar drive belt, and 15×7-inch wheels ranging from heavy-duty steel to ten-hole aluminum. The 1989–93 models are considered the best with their combination of a 225-horsepower fuel-injected 5.0, automatic or five-speed manual transmission, 8.8-inch rear end (first used in 1986), and all available SSP equipment, including a choice of calibrated speedometers (140 or 160 miles per hour), a heater hose inlet restrictor, and an engine oil cooler.

After their service life, the SSP Mustangs were typically sold at auction, with some purchased for drag racing or high-performance street duty. Today, many have been restored to their original highway patrol colors with emergency lights and other equipment.

SPECIAL EDITIONS
ASC/MCLAREN

In 1981, Ford teamed up with McLaren to build a turbocharged Mustang hatchback concept car with air dams and wide-body fenders. The relationship would lead to production ASC/McLaren Mustangs later in the decade.

After building Capris from 1984–86, ASC/McLaren switched to Mustangs in 1987 for a four-year run of two-seater roadsters. Ford embraced the ASC/McLaren Mustang, seeing it as an upscale, Mercedes SL–like vehicle in contrast to the "boy-racer" look of the Mustang GT convertible.

Thanks to Ford's blessing, each ASC/McLaren Mustang started life on the Ford assembly line as a black 5.0 hardtop because it was easier for ASC to chop off the roof than adapt a convertible. Hardtops destined for ASC/McLaren conversion also received convertible chassis bracing and a GT front fascia. At the ASC facility in Livonia, Michigan, the roof was removed from windshield to trunk, additional bracing was added, and the windshield was slanted back 20 degrees for a sleeker appearance. ASC also added the Cambria cloth top that, when lowered, disappeared beneath a hard tonneau cover. After the conversion, which included considerable sheetmetal work at the rear quarters and the addition of the ASC/McLaren ground effects, the cars were stripped and repainted using Sikkens colors in a monochromatic theme. Inside, ASC reconstructed the rear-seat area with Corvette-like storage compartments, covered the factory GT seats with leather, and added a custom console that stretched into the rear compartment area.

By moving to the more popular Mustang, ASC/McLaren nearly tripled convertible sales—for a total of 1,806—during 1987–90. The total could have been more if not for a dispute over licensing and royalty agreements. Only sixty-five were sold in 1990, the ASC/McLaren Mustang's final year.

SPECIAL EDITIONS
BULLITT

FUN FACT
Although Highland Green mimicked the color of Steve McQueen's 1968 movie car, the 2001 Bullitt GT was also offered in True Blue and Black.

BY THE NUMBERS
6,500: Number of limited-production 2001 Bullitt GTs ordered. Only 5,582 made it through the scheduled production run.

HISTORICAL TIDBIT
The popularity of the 2001 Bullitt GT inspired an encore in 2008 and 2009 in Highland Green or Black, followed in 2018 by an S550 Bullitt model to commemorate the movie's 50th anniversary.

In 2001, with the New Edge–styled SN-95 Mustang in its third year and no plans for another styling update until a scheduled makeover in 2005, the Mustang needed a shot of adrenaline to spur interest and sales. Under chief engineer Art Hyde, Team Mustang was searching for a special edition when it fell right into their laps. For the Los Angeles Auto Show, Mustang design manager Sean Tant created a Mustang themed around the 1968 fastback driven by Steve McQueen in *Bullitt*, a 1968 movie known for its San Francisco chase scene. When Tant's Highland Green car drew rave reviews, Hyde and Team Mustang knew they had found their special edition.

For the 2001 Bullitt GT, Team Mustang could have simply added Highland Green paint and vintage-style five-spoke wheels. However, it took the concept much further, tapping into the 1960s vibe with vintage-like instrument graphics and shift ball, aluminum pedal covers, and a brushed aluminum fuel filler cover. Even the C pillar, quarter panel moldings, and rocker panels were modified to create more of a vintage look.

Unlike previous Mustang special editions that were primarily appearance packages, the Bullitt GT package also added performance enhancements, starting with the ¾-inch lower suspension with Tokico struts and shocks, unique stabilizer bars, and subframe connectors. Brembo supplied the 13-inch front brake rotors with the first-time use of red calipers. By utilizing a better flowing aluminum intake manifold with twin-inlet throttle body, high-flow mufflers, and underdrive accessory pulleys, horsepower was increased by 10 over the standard GT.

The 2001 Bullitt GT generated much-needed buzz for the Fox-4 Mustang in its waning years.

SPECIAL EDITIONS
MACH 1

KEY PERSON

Scott Hoag, Team Mustang customization engineer, spearheaded the creation of the 2003–04 Mach 1 as a follow-up to the 2001 Bullitt GT special edition.

BY THE NUMBERS

13.13 seconds: Quarter-mile elapsed time posted with a factory stock 2003 Mach 1 by *Muscle Mustangs & Fast Fords* magazine, quicker than the original 1969 Mach 1s with the 428 Cobra Jet

TOP QUOTE

"Modern engines are well-tuned and balanced, so there's really no 'shake' with a Shaker hood. But Team Mustang is full of enthusiasts who know what it's like to pull a Mach 1 up next to a Chevelle SS with cowl induction and show off."
—Scott Hoag

The popularity and success of the 2001 Bullitt GT sent Team Mustang searching for a follow-up. This time, customization engineer Scott Hoag wanted to bring back the legendary Mach 1 name from the 1960s. Noting that Classic Design Concepts offered an aftermarket "Shaker" hood scoop nearly identical to the Ram-Air system from 1969–70, Hoag envisioned the through-the-hood scoop as the perfect styling component for a modern Mach 1.

Hoag credits "car guys" like Art Hyde and Ford vice president Chris Theodore with pushing the Mach 1 program through. Like the Bullitt GT, it was more than just a visual package. With the 2003 Cobra equipped with a supercharged version of the 4.6-liter DOHC engine, Team Mustang slipped the Mach 1 between the GT and Cobra models by bringing back the naturally aspirated DOHC engine with 305 horsepower. On top was an exact duplicate of the 1969–70 Shaker scoop with ducting underneath to funnel cooler outside air into the factory air cleaner. The Mach 1 was available with both five-speed manual and four-speed automatic transmissions, both feeding into 3.55:1 rear gears.

In addition to the Shaker scoop, a number of other styling cues tied the new Mach 1 to the original from 1969, including a black hood and side stripes, chin spoiler, pedestal-mount rear spoiler, and new 17-inch "Heritage" wheels that resembled vintage Magnum 500s. Suspension mods lowered the Mach by ½ inch, and subframe connectors stiffened the chassis. Inside, appointments such as ribbed "comfort weave" vinyl seats, a Bullitt-style instrument panel, and an aluminum shift ball added to the vintage appeal.

Like the 2001 Bullitt GT, the 2003 Mach 1 was enthusiastically welcomed by the Mustang community. Ford planned to limit availability to 6,500, but nearly ten thousand were sold for 2003. The Mach 1 returned as a carryover 2004 model available in two new exterior colors.

ANATOMY OF MUSTANG MARKETING

Early on, focus group studies predicted that the 1965 Mustang would be a sales success. Bolstered by the encouraging reports, Ford cranked up its marketing efforts to get the word out about the new sporty-looking compact with a low sticker price. From TV, magazine, and newspaper advertising to launching at the New York World's Fair, the extent and expense of the 1965 Mustang's marketing campaign was unprecedented in US automotive history.

For the Mustang's first birthday, Ford turned to packaged equipment groups—GT and Décor interior—to spur Mustang sales. A year later, in the spring of 1966, Ford reached into its marketing box of tricks to create the low-priced Sprint 200, a six-cylinder Mustang with a group of added options. Springtime sales promotions would be used throughout the Mustang's first generation, nationally as the Sprint or Grabber (1970), combined with Ford sales districts often creating their own promotions based on special paint colors.

During the Mustang II era, Ford saw the opportunity to reignite a performance image for the Mustang with the 1976 Cobra II, then hit a marketing home run by supplying a white-with-blue-stripes model to the hit TV series *Charlie's Angels* as the vehicle of choice for popular actress Farrah Fawcett.

In more recent years, special editions spawned interest during the waning years of Mustang generations. As the aging Fox-body neared its end, Ford introduced special "Feature Car" convertibles in monochromatic red, white, and yellow. Later, the 2001 Bullitt GT and 2003–04 Mach 1 created excitement and additional sales during the final years of the SN-95 era.

GLOSSARY: SPECIAL EDITIONS

ASC: American Specialty Cars, a.k.a. American Sunroof Corporation, a company that specialized in aftermarket roof and body systems, including convertible conversions and moonroofs. The company produced the two-seater ASC/McLaren Mustang roadster from 1987 to 1990.

Brembo: An aftermarket manufacturer of high-performance brakes used on a number of Mustang special editions.

Bullitt: A 1968 movie starring Steve McQueen and a Highland Green Mustang fastback, known for its chase scene through San Francisco. Ford paid tribute to the movie car by producing Bullitt Mustang models in 2001, 2008–09, and 2018.

Feature car: Special monochromatic red, white, and yellow Mustang convertibles produced in 1992–93.

Grabber: The name for Ford's brighter paint colors in the early 1970s, highlighted by a special-edition Mustang SportsRoof in 1970.

Olympic Games: International sporting events, winter and summer, held every four years. Ford honored the 1972 US Olympic teams by creating special white, red, and blue Sprint models, including Mustangs, during the 1972 model year.

Pace Car White: A special code-C shade of white paint used for the 1964 ½ Mustang pace car replica hardtops.

SSP: The Special Service Package, created as an equipment group to prepare 1983 to 1993 5.0-liter Mustangs for high-speed pursuit and other strenuous duties when used by police departments and Highway Patrols.

RACING

RACING
EUROPEAN RALLIES

KEY PERSON

Alan Mann, who took Mustangs supplied by Ford and partially prepared by Holman Moody, undertook the additional preparation needed to make them rally winners in Europe.

FUN FACT

Alan Mann brought a fourth Mustang to the Tour de France rally. He used the red hardtop as personal transportation until parts were scavenged to repair the race cars.

HISTORICAL TIDBIT

To commemorate the 50th anniversary of the Mustang's first competition victory, Holman Moody offered a limited-edition 2014 Mustang TDF—for "Tour de France."

In America, marketing the 1965 Mustang to the public took priority over racing. Overseas, however, the Mustang needed to earn respect by competing in European road rallies. When Ford built the earliest preproduction Mustangs in February 1964 for the New York World's Fair and other promotional purposes, several of those first cars were shipped to Britain's Alan Mann Racing for testing as rally cars.

Mann's race team was uniquely suited to prepare Mustangs for the grueling, multiday European rally events. In 1962, Mann campaigned Ford Zephyrs and Cortinas, catching the attention of Ford's Holman Moody race shop. For 1964, Mann landed the contract to race Falcons and Mustangs. Notably, Mann would enter eight Falcons at Monte Carlo, finishing first in class and second overall.

Mann also took delivery of six Mustang hardtops for rally preparation, which included replacing the original engines with 289s prepared by Holman Moody. Two of the cars entered Belgium's 6,000-kilometer Liège-Sofia-Liège Rally in August 1964, representing the first time the Mustang saw serious competition. Unfortunately, both cars crashed.

The following month, Mann arrived in Lille, France, with a trio of red Mustang hardtops for the Touring class in the Tour de France Automobile rally, a seventeen-stage race that started in Lille and ran through the French towns of Reims, Le Mans, Monza, and Pau. At the end of the grueling ten-day competition, Mustang co-drivers Peter Procter and Andrew Cowan took the checkered flag for a first-place finish in class and eighth overall. It would go down in history as the first professional race win for Ford's new Mustang.

RACING
SCCA B-PRODUCTION

KEY PERSON
Jerry Titus, Shelby team driver, followed up his SCCA Pacific Coast B-Production title with a national championship in a competition 1965 GT350.

HISTORICAL TIDBIT
The year 1965 was the only year for a two-seat GT350 and competition model. After winning the B-Production national championship, attention turned to selling more street cars for 1966.

FUN FACT
Today, the factory-built GT350 race cars are commonly called R models. However, that term did not appear until the 1970s as enthusiasts looked for a way to differentiate the street and race models. Shelby American always referred to them as "competition models."

When the Thunderbird abandoned its original two-seater configuration to become a four-seat luxury model in 1958, Ford no longer had direct competition for Chevrolet's Corvette. While the Cobra stuck a bandage over the situation starting in 1962, Shelby American was not capable of building enough of its hybrid two-seaters to match Chevy's production capabilities for the Corvette. With the 1965 Mustang, Ford recognized an opportunity to compete with Corvette by transforming the fastback into a "sports car" for Sports Car Club of America road racing.

Not surprisingly, Ford tasked Carroll Shelby with the job. Shelby called John Bishop, SCCA's executive director, and learned that the Mustang would fit the sanctioning body's definition of a sports car after removing the rear seat, installing larger brakes, improving the suspension, and increasing horsepower to at least 300. With those changes, the Mustang would be classified in the SCCA's B-Production class, which included the Jaguar XKE, Sunbeam Tiger, Lotus Elan, and—most importantly—Corvette.

By January 1965, Shelby had the required one hundred two-seat GT350s built and ready for SCCA inspection. A number of those cars were competition models equipped with a race-prepared 289 High Performance engine, 32-gallon fuel tank, stripped interior, and a plexiglass rear window with a 2-inch opening at the top to vent air from inside the car.

In 1965, Shelby GT350s won the B-Production class in every SCCA region except one. At November's American Road Race of Champions, which would determine the national champion, eight of the qualified fourteen entries were GT350s. Shelby driver Jerry Titus won the race, with Bob Johnson second in his GT350, to capture the B-Production national championship for Mustang.

The SCCA created the Trans-American Sedan Championship in 1966. There were two classes: Group I with under-2.0-liter engines for predominantly European cars and Group II for cars with over-2.0-liter (305-cubic-inch) engines, a class that was perfect for the emerging American pony car segment that included Mustang and Barracuda. Camaro and Firebird would join the fray in 1967, followed by Challenger, Barracuda, and Javelin in 1970. The Trans-Am glory years were 1966 to 1972, a time when manufacturers built special models, including Ford's Boss 302 Mustang, to homologate engines and race parts.

Trans-Am rules required four seats, making the Shelby GT350 ineligible because it had been homologated as a two-seater. So Ford asked Shelby American to prepare a batch of 1966 Mustang hardtops modified to the GT350 competition specifications. These "Group II" Mustangs battled the Plymouth Barracudas and Dodge Darts throughout the 1966 season. Shelby driver Jerry Titus won the final race at Riverside to overtake Dodge and deliver the 1966 Trans-Am manufacturer's championship to Ford.

Mustang would be a dominant force during Trans-Am's first seven years, winning three manufacturer's championships and finishing second in the other four seasons.

In 1967, Mustang won the championship in the hands of Shelby American, which campaigned a yellow "Terlingua" hardtop driven by Titus. After Penske-prepared Camaro Z/28 domination in 1968 and 1969, Mustang reclaimed the Trans-Am title in 1970 with Boss 302s prepared by Bud Moore Engineering and driven by Parnelli Jones and George Follmer.

Ford pulled out of racing following the 1970 season but returned to Trans-Am in the 1980s. In 1989, Ford captured the season championship with rookie driver Dorsey Schroeder winning five out of fifteen races in a Roush Mustang. The 1997 Trans-Am season was totally dominated by Roush Mustangs, which won all thirteen races, including eleven straight by driver Tommy Kendall.

RACING
FUNNY CAR

TOP QUOTE
"We could see the handwriting on the wall. Funny Cars were what people wanted to see. They were getting the match races and grabbing the sponsors." —Pat Foster, designer and builder of the Mickey Thompson 1969 Mach 1 Funny Cars

TECHNOLOGICALLY SPEAKING
In recent years, funny cars have been powered by supercharged Chrysler Hemi engines, regardless of the body style.

FUN FACT
Mustang drag racing superstar Gas Ronda was a former dance instructor.

While Mustangs were racking up SCCA road-racing championships from 1965 to 1967, over at the National Hot Rod Association (NHRA), a new class of altered-wheelbase drag cars was gaining popularity. By shifting the center of gravity rearward, more weight was placed on the rear tires for improved traction. By 1964, these ultra-quick and fan-favorite vehicles had front and rear axles moved forward, set-back fuel-injected engines, and long-nose fiberglass bodies, all of which combined for a strange-looking vehicle and the resulting "funny car" description. Over the years, they would evolve into today's Top Fuel drag cars with one-piece, lift-up fiberglass bodies.

The Mustang joined the funny car movement in 1965 with a batch of Hi-Po fastbacks converted by Holman Moody for the NHRA's Factory Experimental class. The following year, the A/FX Mustangs were purpose built with stretched frames and fiberglass bodies, with several distributed to Ford Drag Council members, including Gas Ronda, Hubert Platt, and Dick Brannan.

By 1969, funny car popularity among fans led to the creation of Funny Car classes in both the NHRA and American Hot Rod Association (AHRA). That year, Mickey Thompson led Ford's Funny Car charge with a pair of flip-top Mach 1s built on dragster-style chassis and powered by SOHC 427s. Driver Danny Ongais would win ten of eleven national events.

Along with Top Fuel dragsters, Funny Cars evolved into a top draw for professional drag racing. In 1997, driver John Force switched from Pontiac to Ford and went on to win nine of his fifteen championships in Mustang-bodied funny cars. For 2018, Ford Performance announced a sponsorship for the Mustang funny car driven by Bob Tasca III.

RACING
PRODUCTION RACING

Over the years, Mustang has been put to the test by competing in race series that allowed limited modifications.

Steve Saleen brought Mustang to the forefront of Showroom Stock in the 1980s. In 1987, Saleen campaigned a two-car team with an impressive list of drivers that included Saleen himself, Rick Titus, Desire Wilson, and Lisa Caceres. Trailing Porsche by a few points heading into the 1987 season finale at Sebring, Wilson made a daring move to pass the leading Porsche on the last turn of the last lap, winning the race and securing both driver and manufacturer championships.

In 1995, Ford SVT produced a limited number of SVT Cobra Rs. During the 1996 race season, the 351-powered Mustangs took on the Firebirds in the IMSA Grand Sport class. The Steeda-prepared No. 20 Mustang driven by Boris Said and Shawn Hendricks nabbed the Cobra R's first race victory. After a season-long battle with the Firebirds, the Cobra R teams came up one point short of taking the manufacturer's championship.

By 1990, the Showroom Stock series had evolved into the World Challenge. In 1995, Saleen made news by forming the Saleen/Allen "RRR" Speedlab team with Tim Allen, the popular comedian and actor from the TV series *Home Improvement*. Saleen and Allen co-drove to win their first race at the end of the 1995 season, then went on to secure World Challenge manufacturer championships for Ford in 1996, 1997, and 1998.

Looking for a race heritage for the 2012 Boss 302 Mustang, Ford Racing supplied race-ready Boss 302S Mustangs for World Challenge and other road-racing series. During the first year of competition for the Boss 302S, driver Paul Brown nabbed five victories and three second-place finishes in the twelve races, dominating the season and nailing down the 2011 manufacturer's championship for Ford.

RACING
5.0 DRAG RACING

FUN FACT
Ford engineer Brian Wolfe was one of the original 5.0 Mustang racers in his naturally aspirated 1986 Mustang GT. He would be named director of Ford Racing in 2008.

BY THE NUMBERS
7.79 at 185 miles per hour: Quarter-mile time in 1995 by "Racin'" Jason Betwarda's 1987 Mustang GT convertible, which was known as the "World's Fastest 5.0" in 1996

KEY PERSON
"Stormin'" Norman Gray, the New York City rock-and-roll tour jacket manufacturer, made the first ten-second pass in a Roush-built 5.0. He would spur the charge into organized Pro 5.0 racing.

Inexpensive, lightweight, fast from the factory, and easily modified with aftermarket speed parts, the 5.0-liter Mustang ruled US streets during the late 1980s and 1990s. Soon, red-light-to-red-light bragging rights weren't enough, leading racers to sanctioned drag strips to prove their Mustang's capability with time slips. *Muscle Mustangs and Fast Fords* editor Steve Collison reported on his 13-second "Mean Mr. Mustang" project, while *Super Ford* put together an invitational race that pitted a variety of modified Mustangs—naturally aspirated, boosted with nitrous oxide, supercharged, and turbocharged—on the same track on the same day. The publicity spawned even more enthusiasm as builders/drivers such as Brian Wolfe, Gene Deputy, "Stormin'" Norman Gray, and Pete Misinsky became national heroes in their quest for more power and quicker elapsed times from their 5.0 Mustangs.

For Bill Alexander and Gary Carter, the timing couldn't have been better when they founded their Fun Ford Weekend (FFW) in 1989 as a quarter-mile venue for Ford drag racers. FFW would soon create the first true classes for heads-up 5.0 drag racing, from grassroots categories to the wild, wheel-standing Pro 5.0. The popularity inspired the creation of the National Mustang Racers Association (NMRA) and the World Ford Challenge, designed as the "Super Bowl" of Mustang drag racing.

As Pro 5.0 racers found more power and speed through combinations of twin turbocharging and nitrous, leading to sub-9-second quarter-mile elapsed times at over 180 miles per hour, the sanctioning bodies began allowing wider rear slicks and tube-frame chassis setups in the interest of safety. After Ford's switch to the 4.6 modular engine for the 1996 Mustang, the strictly 5.0 classes slowly disappeared as they evolved into Outlaw and Renegade categories.

MOST SUCCESSFUL DRIVERS

During the Mustang's fifty-plus-year history, a number of drivers have forever linked their names to the Mustang's competition legacy.

Jerry Titus: As a team driver for Shelby American, Titus captured the 1965 B-Production national championship in a GT350, then won the final Trans-Am race of 1966 in a Shelby-prepared hardtop to give Mustang the points needed to win the season championship. The following year, Titus drove a 1967 Mustang hardtop, campaigned by Shelby, to the 1967 Trans-Am championship.

Al Joniec: Joniec campaigned numerous Mustangs, including his popular "Batcar" 1965 fastback, during his drag racing career, but he is also known for winning the 1968 NHRA Winternationals Super Stock championship in the debut of the 1968 Cobra Jet Mustang.

Parnelli Jones: Jones was already recognized as one of the world's top road racers when he joined Bud Moore Engineering in 1969 to drive a Boss 302 Mustang in Trans-Am. He and teammate George Follmer would capture the 1970 Trans-Am championship, forever linking Jones to his No. 15 School Bus Yellow Mustang.

Steve Saleen: Saleen was instrumental as a manufacturer and driver in the Mustang's rise to the top of Showroom Stock racing, winning several 1987 Escort Endurance series championships with co-driver Rick Titus.

Dorsey Schroeder: Schroeder won five out of fifteen races in a Roush Racing Mustang to win the Trans-Am driver's championship in 1989, the Mustang's twenty-fifth anniversary year.

Tommy Kendall: Driving for Roush Racing, Kendall took the checkered flag at eleven of thirteen Trans-Am races in 1997, as Roush Mustangs won every race that year.

John Force: As one of the top personalities in professional drag racing, Force put Mustang at the top of the Funny Car class by winning nine of his fifteen championships with Mustang bodies.

GLOSSARY: RACING

B-Production: A popular SCCA road-racing class in the 1960s for production two-seat sports cars, including Shelby GT350s and Corvettes.

FR500CJ: The part number for a series of turn-key Mustang drag cars from Ford Racing Performance Parts. Also known as Cobra Jets, the CJ Mustangs came from Ford with a race-prepped supercharged engine, special drag racing suspension, 9-inch rear axle, and roll cage.

Funny Car: The description coined for full-bodied, altered-wheelbase drag cars in the 1960s, which evolved into today's NHRA Funny Car class with supercharged, nitromethane-fueled Hemi engines and lift-off fiberglass or carbon-fiber bodies.

Homologation: By definition, "to approve or confirm officially." In racing, most sanctioning bodies required that special equipment (engines, suspension components, spoilers, etc.) be available on production cars. Generally, a set number of cars (100, 500, 1,000, and so on) had to be built and sold to the public, resulting in Mustangs like the 1965 Shelby GT350 and 1969–70 Boss 302/429s.

Pro 5.0: The top class of 5.0-liter Mustang drag racing during the 1980s and 1990s. It evolved from grassroots street-racing activities, typically with supercharger, turbocharger, or nitrous-oxide power-adders.

Rally, Rallye (European): A type of road racing, popular in Europe in the 1960s, typically run in segments on public roads and highways, sometimes over the course of several days.

Trans-Am: Short for the SCCA's Trans-American Sedan Championship series at race venues across North America. Created in 1966, Trans-Am originally incorporated modified production sedans, including pony cars like the Mustang, and evolved into purpose-built tube frame cars.

Super Stock: A drag-racing class for stock vehicles with approved modifications

INDEX